# Loft Conversions

# Loft Conversions

## SECOND EDITION

## John Coutts

MA (Oxon)

**WILEY-BLACKWELL**

A John Wiley & Sons, Ltd., Publication

This edition first published 2013
© 2006 John Coutts
© 2013 John Coutts

Blackwell Publishing was acquired by John Wiley & Sons in February 2007. Blackwell's publishing program has been merged with Wiley's global Scientific, Technical and Medical business to form Wiley-Blackwell.

*Registered Office*
John Wiley & Sons, Ltd, The Atrium, Southern Gate, Chichester, West Sussex, PO19 8SQ, UK

*Editorial Offices*
9600 Garsington Road, Oxford, OX4 2DQ, UK
The Atrium, Southern Gate, Chichester, West Sussex, PO19 8SQ, UK
2121 State Avenue, Ames, Iowa 50014-8300, USA

For details of our global editorial offices, for customer services and for information about how to apply for permission to reuse the copyright material in this book please see our website at www.wiley.com/wiley-blackwell.

*Library of Congress Cataloging-in-Publication Data*

Coutts, John, 1965–
    Loft conversions / John Coutts. – 2nd ed.
        p.   cm.
    Includes bibliographical references and index.
    ISBN 978-1-118-40004-3 (pbk. : alk. paper)   1. Lofts–Remodeling for other use.   I. Title.
    TH3000.L63C68 2012
    728′.314–dc23

                                    2012009293

A catalogue record for this book is available from the British Library.

Wiley also publishes its books in a variety of electronic formats. Some content that appears in print may not be available in electronic books.

Cover image courtesy of FreshPaint/www.Bigstock.com
Cover design by Meaden Creative

Set in 10/12.5pt Minion by SPi Publisher Services, Pondicherry, India
Printed and bound in Malaysia by Vivar Printing Sdn Bhd

1   2013

# Contents

*A colour plate section falls between pages 162 and 163*

# Preface

The purpose of this book is to provide technical, regulatory and practical guidance on loft conversions in single-family dwellings. It is the most comprehensive book of its sort and is the result of extensive research and consultation with regulatory bodies and practitioners.

Since the publication of the first edition in 2006, *Loft conversions* has become established as the definitive source of guidance for architects, builders, surveyors and others professionally involved in the process of loft conversions.

This extensively-updated second edition takes into account significant changes to Building Regulations and planning law that have taken place since the first edition was published. It also contains a new section on sustainability and zero carbon approaches to loft conversions.

*John Coutts*
*October 2012*

# Acknowledgements

I would like to thank the following organisations and businesses for their invaluable co-operation and assistance in the preparation of this book: the Building Research Establishment (BRE), British Woodworking Federation (BWF), British Standards Institution (BSI), Cooper & Turner Ltd, Coopers Fire Ltd, Department for Communities and Local Government (DCLG), Denmay Steel, Economic and Social Data Service (ESDS), Energy Saving Trust (EST), English Heritage, Euroform Products Ltd, Federation of Master Builders (FMB), Green Structures, Local Authority Building Control (LABC), Polytank Group Ltd, South London Lofts Ltd, TRADA Technology Ltd, Trussed Rafter Association (TRA) and the Welsh Assembly Government.

Material reproduced from the Approved Documents and other government sources is Crown copyright and is reproduced with permission of the Controller of the HMSO.

# 1 Planning and legal considerations

This chapter examines the influence of planning and other legal mechanisms on the loft conversion process in England. Obligations imposed by the Building Regulations are considered in Chapter 2.

The controls and mechanisms examined both here and in Chapter 2 are largely separate from each other. Planning and building control, for example, are administered independently. Approvals granted under one mechanism do not automatically confer rights under another, nor are they intended to. Building Regulations and planning law have specific and generally unrelated aims.

## PERMITTED DEVELOPMENT

Most loft conversions are carried out under permitted development legislation. Where permitted development rights exist, no specific application for planning permission is required, provided that work is carried out in accordance with the legislation. Permitted development rights apply to dwellinghouses only. A loft conversion in a building containing one or more flats, or a flat contained within such a building, would require planning permission. The following section considers current permitted development legislation for England only.

## Permitted development law

Permitted development legislation is set out in The Town and Country Planning (General Permitted Development) (Amendment) (No. 2) (England) Order 2008. This came into force on 1 October 2008 and represents the first major change in planning law relevant to small-scale domestic building works, such as loft conversions, since 1995.

One of the notable features of the 2008 General Permitted Development Order (GPDO 2008) is that it is rather more generous in its scope than the earlier legislation. It dispenses with the principle of a whole-dwelling volume allowance (at least as far as loft conversions are concerned) and only the volume of the roof is now considered (Fig. 1.1). A ground floor extension to a dwellinghouse, whether proposed or existing, no longer counts against a loft conversion.

*Loft Conversions*, Second Edition. John Coutts.
© 2013 John Coutts. Published 2013 by Blackwell Publishing Ltd.

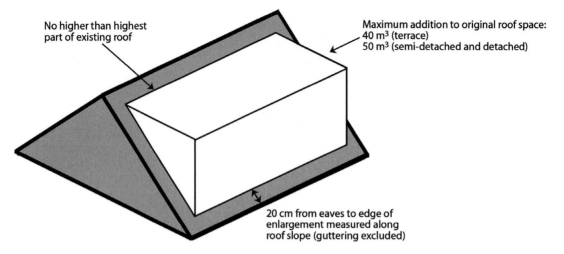

**Fig. 1.1**   Permitted development (England): primary constraints.

Reproduced below are three extracts from The Town and Country Planning (General Permitted Development) (Amendment) (No. 2) (England) Order 2008. All are relevant, or potentially relevant, to loft conversions. The meaning and implications of the GPDO 2008 are considered in the next section.

## Class B

### Permitted development

B.   *The enlargement of a dwellinghouse consisting of an addition or alteration to its roof.*

### Development not permitted

B.1  *Development is not permitted by Class B if –*
   *(a)  any part of the dwellinghouse would, as a result of the works, exceed the height of the highest part of the existing roof;*
   *(b)  any part of the dwellinghouse would, as a result of the works, extend beyond the plane of any existing roof slope which forms the principal elevation of the dwellinghouse and fronts a highway;*
   *(c)  the cubic content of the resulting roof space would exceed the cubic content of the original roof space by more than –*
      *(i)  40 cubic metres in the case of a terrace house, or*
      *(ii)  50 cubic metres in any other case;*
   *(d)  it would consist of or include –*
      *(i)  the construction or provision of a veranda, balcony or raised platform, or*
      *(ii)  the installation, alteration or replacement of a chimney, flue or soil and vent pipe; or*
   *(e)  the dwellinghouse is on article 1(5) land.*

## Conditions

B.2 *Development is permitted by Class B subject to the following conditions –*

    (a) *the materials used in any exterior work shall be of a similar appearance to those used in the construction of the exterior of the existing dwellinghouse;*

    (b) *other than in the case of a hip-to-gable enlargement, the edge of the enlargement closest to the eaves of the original roof shall, so far as practicable, be not less than 20 centimetres from the eaves of the original roof; and*

    (c) *any window inserted on a wall or roof slope forming a side elevation of the dwelling-house shall be –*

        (i) *obscure-glazed, and*

        (ii) *non-opening unless the parts of the window which can be opened are more than 1.7 metres above the floor of the room in which the window is installed.*

## Interpretation of Class B

B.3 *For the purposes of Class B 'resulting roof space' means the roof space as enlarged, taking into account any enlargement to the original roof space, whether permitted by this Class or not.*

# Class C

## Permitted development

C. *Any other alteration to the roof of a dwellinghouse.*

## Development not permitted

C.1 *Development is not permitted by Class C if –*

    (a) *the alteration would protrude more than 150 millimetres beyond the plane of the slope of the original roof when measured from the perpendicular with the external surface of the original roof;*

    (b) *it would result in the highest part of the alteration being higher than the highest part of the original roof; or*

    (c) *it would consist of or include –*

        (i) *the installation, alteration or replacement of a chimney, flue or soil and vent pipe, or*

        (ii) *the installation, alteration or replacement of solar photovoltaics or solar thermal equipment.*

## Conditions

C.2 *Development is permitted by Class C subject to the condition that any window located on a roof slope forming a side elevation of the dwellinghouse shall be –*

    (a) *obscure-glazed; and*

    (b) *non-opening unless the parts of the window which can be opened are more than 1.7 metres above the floor of the room in which the window is installed.*

### *Class G*

*Permitted development*

G. *The installation, alteration or replacement of a chimney, flue or soil and vent pipe on a dwellinghouse.*

*Development not permitted*

G.1  *Development is not permitted by Class G if –*
    (a)  *the height of the chimney, flue or soil and vent pipe would exceed the highest part of the roof by 1 metre or more; or*
    (b)  *in the case of a dwellinghouse on article 1(5) land, the chimney, flue or soil and vent pipe would be installed on a wall or roof slope which –*
        (i)  *fronts a highway, and*
        (ii)  *forms either the principal elevation or a side elevation of the dwellinghouse.*

## Commentary on permitted development provisions – England

A degree of caution should be exercised when exercising rights associated with permitted development. Where any doubt exists, clarification should be sought from the local planning authority and a Lawful Development Certificate obtained (see p. 13) before undertaking any work.

Permitted development rights are not universal: they do not apply to flats, for example, nor do they apply to dwellinghouses on designated land (see section on Article 1(5) land, below). It is also emphasised that development that is not permitted under one class may be permitted development under another: chimneys, soil pipes and solar panels all fall into this category.

It should also be noted that interpretation of the GPDO varies considerably between local planning authorities. Areas of inconsistency include:

- Raising a party wall (see also appeal decision letter in Appendix C)
- Providing a highway-facing roof window in a dwelling in a conservation area

There are also risks when working at the volume limits of permitted development. A local planning authority has discretionary powers to take enforcement action if, in its view, there is an unacceptable breach of planning control. In cases where any degree of doubt exists, therefore, it is prudent to consult the local planning authority before work commences.

The Department for Communities and Local Government has sought to clarify some of the 2008 provisions and has published two supporting documents. These are: *Changes to Householder Permitted Development 1 October 2008 – Informal Views from Communities and Local Government* (this document has now been superseded) and *Permitted development for householders – Technical guidance* (published August 2010). The latter document is described as 'CLG guidance' where it is referenced below.

The following section highlights areas that require consideration in the 2008 GPDO.

## Development within the curtilage of a dwellinghouse

This is the title of Part 1 of the 2008 GPDO. The meaning of 'curtilage' is subject to a degree of interpretation. This has important implications for conversions that involve raising a party wall between dwellinghouses: some local authorities consider raising a party wall to be permitted development, others do not. See *Curtilage: raising party walls*, below, and Appendix C.

### B. Dwellinghouse

The GPDO 1995 definition remains valid in this section for the purposes of Part 1:

'dwellinghouse' does not include a building containing one or more flats, or a flat contained within such a building.

### B.1 (a) Defining the highest part of an existing roof

The ridge of a conventional pitched roof is generally its highest part. Where the roof is of slated or tiled construction, it is usually capped with ridge tiles. But the definition of precisely which part of the ridge is to provide the highest point datum remains open to a degree of interpretation, particularly where the original roof is capped with decorative 'crested' ridge tiles which may project more than 150 mm above the apex.

The position with walls and other masonry projections is less ambiguous. In the case of buildings with butterfly roofs and a front parapet wall, local planning authorities have tended historically to take the roof as the highest point, even though the highway-facing parapet is higher (Fig. 3.1c). This position is supported by CLG guidance, which suggests that:

Chimneys, firewalls, parapet walls and other protrusions above the main roof ridge line should not be taken into account when considering the height of the highest part of the roof of the existing house.

### B.1 (b) Roof slopes: principal elevation fronting a highway

Alterations to a roof slope fronting a highway (other than the installation of roof windows in the same plane as the existing roof) are not permitted development. For example, a front dormer window (i.e. one occupying and projecting from the principal roof slope facing a highway) could not be considered permitted development. Planning permission would be needed.

Defining precisely what constitutes a 'principal elevation' is not always a simple matter, however. CLG guidance states the following:

The effect of this [i.e. B.1(b)] is that dormer windows as part of a loft conversion, or any other enlargement of the roof space, are not permitted development on a principal elevation that fronts a highway and will therefore require an application for planning permission. Roof-lights in a loft conversion on a principal elevation may however be permitted development as long as they meet the requirements set out under Class C [].

In most cases, the principal elevation will be that part of the house which faces (directly or at an angle) the main highway serving the house (the main highway will be the one that sets the postcode for the house concerned). It will usually contain the main architectural features such as main bay windows or a porch serving the main entrance to the house. Usually, but not exclusively, the principal elevation will be what is understood to be the front of the house.

There will only be one principal elevation on a house. Where there are two elevations which may have the character of a principal elevation (for example, on a corner plot), a view will need to be taken as to which of these forms the principal elevation.

The principal elevation could include more than one roof slope facing in the same direction – for example, where there are large bay windows on the front elevation, or where there is an 'L' shaped frontage. In such cases, all such roof slopes will form the principal elevation and the line for determining what constitutes 'extends beyond the plane of any existing roof slope' will follow these slopes [].

A highway will usually include public roads (whether adopted or not) as well as public footpaths and bridleways, but would not include private driveways. The extent to which an elevation of a house fronts a highway will depend on factors such as:

(i) the angle between the elevation of the house and the highway. If that angle is more than 45 degrees, then the elevation will not be fronting a highway;
(ii) the distance between the house and the highway – in cases where that distance is substantial, it is unlikely that a building can be said to 'front' the highway. The same may be true where there is a significant intervening area of land in different ownership or use between the boundary of the curtilage of the house concerned and the highway.

### B.1 (c) Cubic content

For a terrace house, the 2008 GPDO allows an addition of up to at $40\,m^3$ and for other types of dwelling (semi-detached and detached), an addition of $50\,m^3$ beyond that of the 'original roof space' (see below). This has the effect of 'capping' the volume of loft conversions but note that these volume figures are limits, not entitlements. In a significant number of cases, it will not be possible to take full advantage of the 'allowance' because the physical footprint of the building, and the limitations imposed by B.1 (a) and B.1 (b), will not permit it.

Note that when the proposed work includes both a hip-to-gable and a dormer conversion, the volumes of both elements must be considered relative to the cubic content 'allowance', that is, both must be deducted from it (Fig. 1.2). Any earlier addition to the cubic content of the original roof space must also be taken into account.

### B.1 (c) Original roof space

The GPDO provides a definition of 'resulting roof space' for the purposes of Class B (i.e. *the roof space as enlarged, taking into account any enlargement to the original roof space, whether permitted by this Class or not*), but it does not define original roof space. However, CLG guidance states that:

'original roof space' will be that roof space in the 'original building' …

in which 'original' means:

… a building as it existed on 1 July 1948 where it was built before that date, and as it was built when built after that date.

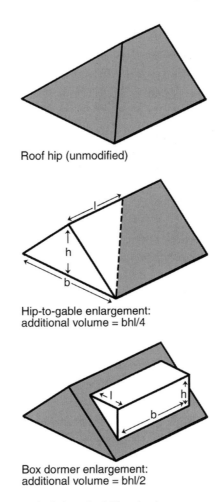

Roof hip (unmodified)

Hip-to-gable enlargement:
additional volume = bhl/4

Box dormer enlargement:
additional volume = bhl/2

**Fig. 1.2**   Roof enlargement: calculation of additional volume.

This is broadly the same definition that is used in the 1995 GPDO.

### B.1 (c) Terrace house

For the purposes of Part 1, the GPDO defines 'terrace house' as follows:

> *... a dwellinghouse situated in a row of three or more dwellinghouses used or designed for use as single dwellings, where –*
> *(a) it shares a party wall with, or has a main wall adjoining the main wall of, the dwellinghouse on either side; or*
> *(b) if it is at the end of a row, it shares a party wall with or has a main wall adjoining the main wall of a dwellinghouse which fulfils the requirements of sub-paragraph (a).*

### B.1 (d)(i) Veranda, balcony or raised platform

Projecting structures (such as balconies) require planning permission. However, a 'Juliet' balcony (with a balustrade but no external platform, and therefore not a true balcony) is normally accepted as permitted development.

### B.1 (d)(ii) Chimney, flue or soil and vent pipe

These are permitted development under Class G, but not Class B.

### B.1 (e) Article 1(5) land

The reference to Article 1(5) land is of importance because it defines where permitted development does not apply and planning permission must be sought. Article 1(5) land (sometimes described as 'designated land') is described in Schedule 1 of the 1995 GPDO and subsequent amendments. Roof extensions are not permitted development in areas that include:

- A National Park
- An area of outstanding natural beauty
- An area designated as a conservation area under section 69 of the Planning (Listed Buildings and Conservation Areas) Act 1990 (designation of conservation areas)
- The Broads
- A World Heritage site

However, these are not the only exceptions. For example, permitted development rights can also be removed by mechanisms including:

- Article IV directions
- Planning conditions
- Listing

### B.2 (a) Materials of similar appearance

The intention is to minimise the visual impact of the conversion and to ensure that it is sympathetic to the existing house. CLG guidance notes that:

> The flat roofs of dormer windows will not normally have any visual impact and so the use of materials such as felt, lead or zinc for flat roofs of dormers will therefore be acceptable.

The face and sides (cheeks) of a dormer window should be finished using materials that are similar in appearance to the existing house:

> … the materials used for facing a dormer should appear to be of similar colour and design to the materials used in the main roof of the house when viewed from ground level. Window frames should also be similar to those in the existing house in terms of their colour and overall shape.

Design guidance published by the local planning authority may provide an indication of what is and what is not likely to be acceptable (see *Sources of planning guidance*, below).

### B.2 (b) Hip-to-gable enlargement

In the majority of cases, this would now be considered to be permitted development.

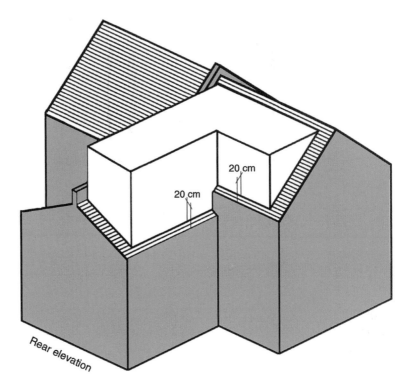

**Fig. 1.3**   L-shaped loft conversion. Typical relationships between conversion, main roof slope, back addition roof slope and eaves.

### B.2 (b) 20 cm from the eaves of the original roof

According to CLG guidance, the 20 cm measurement:

> … should be made along the original roof slope from the outermost edge of the eaves (the edge of the tiles or slates) to the edge of the enlargement. Any guttering that protrudes beyond the roof slope should not be included in this measurement.

The CLG guidance states that the 20 cm setback is required unless it can be demonstrated that it is not practical due to 'practical or structural considerations':

> One circumstance where it will not prove practical to maintain this 20 cm distance will be where a dormer on a side extension of a house joins an existing, or proposed, dormer on the main roof of the house.

Fig. 1.3 illustrates the relationship between an L-shaped loft conversion that encompasses both the principal rear roof slope and the subordinate roof slope of a back addition (outrigger).

### B.2 (c)(i) Obscure glazed

CLG guidance suggests that windows should be obscure glazed to a minimum of level 3 (on a scale of 1 to 5, where 5 represents the highest level of obscuration). One-way glass is not suitable. Note that the scale referred to is an informal one used by glazing manufacturers and suppliers as a guide for their customers. It is not based on a quantifiable standard.

### C.1 (a) 150 mm protrusion

One of the intentions of this is to limit the prominence of protrusions such as roof windows. CLG guidance notes the limitation to project from the roof plane:

> … should not be applied in cases where the roof of an extension to a house that is permitted development under Class A is joined to the roof of the original house. In such cases, the roof of the extension should not be considered as protruding from the original roof.

Note that height considerations in C.1 (a) and C.1 (b) are referenced to the 'original roof' while those in B.1 (a) and B.1 (b) refer to the 'existing roof'.

### C.1 (c)(ii) Solar photovoltaics

Permitted development rules for solar photovoltaics and solar thermal are set out in Part 40 of The Town and Country Planning (General Permitted Development) (Amendment) (England) Order 2008.

## PERMITTED DEVELOPMENT RESTRICTIONS

As noted above, permitted development rights do not apply universally. The major restrictions and ambiguities are considered in more detail below.

## Curtilage: raising party walls

When a full-width loft conversion is carried out in a mid-terrace dwelling, party walls on both sides are sometimes raised at full thickness to form new flank walls. It is emphasised, however, that a considerable number of local planning authorities consider that such conversions fall *outside* the scope of permitted development, arguing that half of each wall lies outside the curtilage of the dwellinghouse. Given the degree of uncertainty surrounding the matter, it would be prudent to ascertain the local planning authority's disposition before undertaking work of this sort. This subject is considered in Appendix C.

## Conservation areas

Permitted development rights are restricted in conservation areas and there is a presumption that any work on a roof in such an area requires planning permission. A new dormer loft conversion, for example, would certainly require an application for planning permission.

Matters are less clear-cut in relation to rooflights and roof windows. Many local planning authorities take the view that these, too, require planning permission by virtue of the provisions of the GPDO.

However, some local planning authorities accept the use of co-planar roof windows (i.e. ones that do not project beyond the plane of the roof), even in roof slopes fronting a highway, under permitted development. As a consequence, roof-space only conversions are sometimes carried out in conservation areas without planning permission.

Because the provisions of the GPDO are open to interpretation, consultation with the local planning authority is advised before undertaking work of *any* description on a roof in a conservation area. The section on planning permission (below) outlines how a planning application may be made and the sources of guidance available.

## Article IV directions

An Article IV direction allows the local planning authority to impose additional controls to restrict work that would normally be permitted development. Under an Article IV direction, such work would require planning permission. There is normally no fee for a planning application that is required as a result of an Article IV direction.

Article IV directions are most commonly used in conservation areas, but not exclusively so, and are generally applied to groups of dwellings rather than individual houses. As far as loft conversions are concerned, an Article IV direction might require that planning permission be sought for the installation of roof windows or replacement roofing materials. But as noted above, such work is often considered to require planning permission anyway by virtue of the provisions of the GPDO.

## Planning conditions affecting permitted development

'Planning conditions' (there is no other specific technical name for them) are used to remove permitted development rights and they are being used increasingly to restrict developments such as loft conversions. Planning conditions are most often applied to new high-density housing, although not exclusively so. Where a local planning authority has applied such a condition, it is necessary to apply for planning permission. Some local authorities, however, waive the normal application fee in such cases.

Because planning conditions may apply to areas and building types that would traditionally have enjoyed permitted development rights, property owners are sometimes unaware of their existence. Conditions such as these ('charges' in legal terms) are recorded in the Local Land Charges register which local authorities are required to maintain. There is, therefore, a case for checking the register even in an area where a building could reasonably be assumed to have permitted development rights.

## Listed buildings

A loft conversion in a listed building would require both planning permission and listed building consent. Current legislation requires consent to be sought for any works to a listed building that would affect its character as a building of special architectural or historic interest. Note that listed building consent is needed for internal alterations as well as external ones, so even a conversion without projecting elements would require consent.

Listed building consent may also be needed for work on buildings within the grounds of a listed building. It is an offence to carry out any work requiring such consent without first obtaining it.

Applications for listed building consent are subject to similar procedures to planning permission, and the process is administered by the local planning authority. There is normally no planning fee for an application for listed building consent.

## OTHER CONDITIONS AFFECTING DEVELOPMENT

### Restrictive covenants

A restrictive covenant imposes conditions on how an owner may use land. This sometimes includes alterations and extensions to buildings. In cases where a restrictive covenant applies, the permission of the original developer may be required before an extension can be built. The restrictive covenant should appear in the land register entries to the title of the property. Note that the register referred to in such cases is maintained by the Land Registry and is quite separate from the Local Land Charges register.

A restrictive covenant can take many forms and can apply to more or less any type of property. It should be noted that local authorities widely apply such covenants to council houses sold under right-to-buy rules and that the local authority therefore has the benefit of the covenant. However, it is generally possible to escape the covenant through negotiation. In cases where there is a very old and apparently out-of-date restrictive covenant, it is sometimes possible to insure against the risk of enforcement. Restrictive covenants are generally a civil matter and operate independently of the planning system.

### Mortgage lenders

Where property is mortgaged, it may be necessary for the householder to obtain bank or building society permission before undertaking a loft conversion. This is because any work might affect the lender's interest in the building and would apply whether or not it had advanced the money to pay for the work. Some lenders may charge a fee for consent.

### Buildings and contents insurance

It is a requirement of most household insurance policies to inform the insurer before undertaking any building alterations; premiums may be adjusted accordingly. This applies to both buildings and contents insurance. In addition, it is generally necessary to modify policies to reflect the greater size of the property once work is complete.

### Tree preservation orders

A tree preservation order (TPO) is an order made by the local authority to protect a tree or group of trees. An application must be made to the local authority to fell or undertake work, including pruning, on a tree subject to a TPO. Such matters are generally dealt with by the local authority's planning department.

### Bats

It is an offence to remove or disturb bats without first notifying the relevant Statutory Nature Conservation Organisation (SNCO) before undertaking any work. In England, this is Natural England and in Wales, the Countryside Council for Wales. Note that bats and their roosts are protected by more than a dozen conventions and sets of legislation including the Wildlife and Countryside Act 1981 and the Conservation (Natural Habitats &c.) Regulations 1994.

## LAWFUL DEVELOPMENT CERTIFICATE

A Lawful Development Certificate (LDC, sometimes called a Certificate of Lawfulness) is a legal document that may be used to establish that proposed building work is lawful and does not require express planning permission. It is of particular use when working at the limits of permitted development rights.

Application for an LDC is made to the local planning authority, which generally bases its decision to grant or refuse on the applicant's submission and drawings. In some cases, the local planning authority will grant a certificate for part of the application. The application should be determined within 8 weeks and a fee is payable in all cases.

The importance of providing accurate drawings and documentation is emphasised in applying for an LDC, and, where a certificate is granted, it is equally important that subsequent work conforms to the drawings submitted. Note that a local planning authority can revoke a certificate issued as a result of false information, or if any material information is withheld.

## PLANNING PERMISSION

An application for planning permission is made to the local planning authority when a proposal cannot be considered under permitted development.

However, before making a formal application and producing detailed drawings, it is prudent to discuss proposals with the local authority's planning department. It is also worth checking planning records to ascertain whether similar conversions have been granted planning permission and to check their planning history.

Note that permitted development rights for highway-facing front dormers were removed by The Town and Country Planning General Development Order 1988.

### Planning applications

Applications may be made either electronically or on paper. Online applications are made using the Planning Portal and details are forwarded automatically to the relevant local authority. Currently, almost half of all applications are submitted online. It is also possible to make payments via the Planning Portal; more than 260 local authorities offer this option.

When applications are made on paper, they are submitted directly to the local authority, generally using a householder application form. Note that the method of submission – electronic or hard copy – does not affect the way a planning decision is made. In both cases, information provided on the application form, and other supporting documents, may be published on the local authority's website.

Once an application has been made, the local authority should decide whether to grant permission (sometimes with conditions) or refuse permission within 8 weeks.

Where permission is refused, granted with conditions that are not acceptable, or if the application is not determined within the statutory period, there is a right of appeal.

Planning permission normally remains valid for 3 years. If work does not start during that time, it may be necessary to re-apply. However, it is possible to extend planning permission before it expires. Note that the local planning authority may carry out checks on compliance during and after construction. Building control surveyors may alert planning enforcement teams to potential planning transgressions.

Note that matters such as restrictive covenants, party wall agreements, questions relating to the Building Regulations and the applicant's personal circumstances are not 'material considerations' as far as deciding a planning application is concerned.

Most of the questions on the householder application form (both on paper and online) are relatively straightforward. However, attention should be given to the following points.

### Ownership

The application must include a completed certificate relating to the ownership of the land. If the proposed conversion involves raising a party wall, the statement of ownership must reflect that the property to which the application relates is not entirely owned by the applicant and there is a requirement to formally notify neighbouring owners about the application before it is submitted. Note that this is not the same as a party wall agreement.

### Location plan (site location plan)

A location plan (generally at 1:1250 scale but see also *Drawings general*) is required that accurately shows the property in relation to roads and other properties. The application site (e.g. the house and garden) must be outlined in red. Any other land owned by the applicant in the vicinity is outlined in blue. The location plan should be based on up-to-date Ordnance Survey mapping and be configured to fit on an A4 page. A north point and a scale are required.

### Site layout plan (block plan)

The site layout plan at 1:500 (or a larger standard scale, e.g. 1:200, but see also *Drawings general*) should indicate the position of the proposed development in relation to the site boundaries and other existing buildings on the site, with written dimensions including those to the boundaries. A north point and a scale are required. Where relevant to the proposed development, the following may also be needed: the position of all buildings, roads and footpaths adjoining the site including access arrangements; all public rights of way crossing or adjoining the site; the position of all trees on the site, and those on adjacent land; the extent and type of any hard surfacing; and boundary treatments including walls or fencing where this is proposed.

### Floor plans and elevations

Plans are drawn to a scale of either 1:50 or 1:100 (but see also *Drawings general*) and should distinguish between existing and proposed structures. Floor plans should show the room layout for the whole building with one drawing for each floor. Doors, windows and wall thicknesses should be indicated. Elevations should show what the proposed conversion will look like from the outside. Where neither side of the conversion is visible, a section drawing should be provided. The drawings must indicate the building materials to be used. Photographs showing aspects of the existing site may be of assistance in making an application.

### Drawings general

All plans and drawings should be to a recognised scale and only metric measurements should be used. Plans should be identified clearly and drawing numbers provided. There is no restriction on the size of drawings. However, drawings sized to A4 or A3 are preferred for online applications.

The shift from analogue to digital drawing has increased the danger of scaling errors, particularly when documents are printed. To reduce this risk, drawings should state the size of paper they are to be printed on and the relevant scale when printed out at that size. As an additional safeguard, a scale bar indicating the length of 1 and 10 m, and written dimensions, should always be included.

### Design and Access Statement

A Design and Access Statement (DAS) is required for a planning application when a dwelling is in a conservation area or a World Heritage site, or is listed.

## SOURCES OF PLANNING GUIDANCE

Because most planning law is essentially negative, local planning authorities are encouraged to produce guidance to clarify what *can* be done. Many local planning authorities now produce detailed guidance on what is acceptable in specific localities.

However, changes to the planning system (and the ease with which documents can now be produced) mean that the volume of statutory and non-statutory guidance has proliferated in recent years, so it is not always easy to unearth relevant documentation.

When searching for potentially useful information, it is worth noting that the words 'loft conversion' are often not used in documentation published by local planning authorities. The following expressions are more commonly used instead:

- Roof extension
- Roof alteration
- Loft extension

Advice on relevant documentation, and whether or not its use is a material consideration for the purposes of an application, should be sought from the local planning authority. It is emphasised that considerable time and effort can be conserved at the proposal stage if relevant local guidance is taken into account.

When producing drawings for an application for planning permission, the following types of documents may provide detailed guidance, or contain references to where relevant guidance might be found.

## Supplementary planning guidance

Most local planning authorities produce supplementary planning guidance (SPG) that expands on statutory policies. In some cases, this will include detailed guidance on residential extensions such as loft conversions. Such guidance, while non-statutory in itself, is taken into account as a material planning consideration when determining applications. The weight accorded to SPG increases if it is prepared in consultation with the public and has been the subject of a council resolution.

## Supplementary planning documents

Site-specific guidance on residential extensions and alterations is sometimes included in supplementary planning documents (SPDs).

## Design guides

These guides are usually drawn up for conservation areas, often after consultation with residents. In the case of loft conversions, a design guide might provide indications of acceptable scale and use of materials in a proposed dormer construction. Where adopted, design guides have the same status as supplementary planning guidance and are thus taken into account as a material planning consideration.

## Design codes

A design code provides illustrated design rules and requirements which instruct and may advise on the physical development of a site or area. These generally paint a broader picture than design guides (see above), but may contain potentially useful references to preferred forms and choice of materials.

## Local Development Framework

This encompasses all the local planning authority's local development documents, including Development Plan documents and supplementary planning documents. The Local Development Framework (LDF) replaces the Unitary Development Plan.

## Unitary Development Plan

The Unitary Development Plan (UDP) is now replaced by the Development Plan system, although through transitional provisions they will continue to operate in many cases. The UDP may set out acceptable roof alterations in some detail.

## THE PARTY WALL ETC. ACT 1996

The Act usually applies when lofts are converted and it provides a framework for preventing and resolving disputes in relation to party walls. Anyone planning to carry out work of the kinds described in the Act must give notice of their intentions to the adjoining owner. The Act applies in England and Wales and is invoked by the building owner. Local authorities are not usually involved in the process.

In semi-detached and terraced dwellings, it is usually necessary to carry out work on a wall shared with another property (a party wall) as part of a loft conversion. Any work on a party wall, other than minor operations, is likely to fall within the scope of the Party Wall etc. Act 1996. Some common examples of work on party walls as part of a loft conversion include:

- Cutting into an existing party wall to provide support for a beam
- Constructing a dormer stud cheek over part of an existing party wall
- Raising the height of an existing party wall to form a new flank gable wall
- Raising a compartment wall if there is no separation in the roof void

Section 2 of the Act deals with work to existing party walls and, if the correct procedures are followed, it confers a number of valuable rights, including rights of access (subject to conditions). It is emphasised that the Act should be invoked even if the proposed work does *not* extend beyond the centreline of the party wall. For example, the Act would still apply where a beam with a bearing of 100 mm is to be supported by a 9" wall (i.e. penetrating less than half the thickness of the wall).

Perhaps the single most important factor in preventing disputes from arising is the building owner's relationship with the adjoining owner. Wherever possible, the building owner should discuss the proposed work with the adjoining owner before notice is served.

## Procedure

The building owner must serve a *party structure notice* on any adjoining owner. A party structure notice must be served at least 2 months before the planned starting date for work to the party wall. Note that the notice is only valid for 1 year. There is no official form for giving notice, but the Act stipulates that the notice contains the following:

- The name and address of the building owner (joint owners must all be named)
- The nature and particulars of the proposed work (drawings may be included)
- The date on which the proposed work will begin

Although not specifically mentioned in the Act, it might be prudent also to include in the notice:

- The address of the building to be worked on
- The date of the notice
- A statement that it is a notice under the provisions of the Party Wall etc. Act 1996

The adjoining owner's agreement and written consent to the proposed work, if it is received, does not relieve the building owner of obligations under the Act – for example, the requirement to avoid unnecessary inconvenience while work is carried out. Neither does it eliminate the possibility of differences arising between the adjoining owner and the building owner once work has begun. The two-month period between serving notice and the planned commencement of work may be reduced with the agreement of the adjoining owner.

## Disputes

If the adjoining owner disagrees and does not consent to the work proposed, one of two approaches may be adopted:

- The building owner and adjoining owner concur in the appointment of a single surveyor – the 'agreed surveyor'. The agreed surveyor produces an 'award' (sometimes called a party wall award) that sets out in detail the work proposed and conditions attached to it.
- The building owner and adjoining owner each appoint a surveyor to settle any differences and to produce an award as outlined above. In such cases, the appointed surveyors select an additional surveyor – the 'third surveyor' – who would be called in to mediate if the two appointed surveyors cannot reach an agreement.

Note that if the adjoining owner fails to respond to the party structure notice within 14 days of service, a dispute is considered to have arisen. In this case, a surveyor is appointed on behalf of the adjoining owner and the procedure described above is followed.

The Act defines a surveyor as a person who is not party to the matter. This means, in theory at least, that anybody can act as a surveyor in a party wall dispute. In practice, of course, it would be prudent to appoint, or agree on the appointment of, a person with a good knowledge both of construction and of administering the Act. Surveyors appointed under the dispute resolution procedure of the Act must consider the interests and rights of both the building owner and the adjoining owner. They do not act as advocates for the respective owners, and an award must be drawn up impartially. Fees for the surveyor (or surveyors) are generally paid by the building owner.

Details of surveyors specialising in the Party Wall etc. Act 1996 may be obtained from the Royal Institution of Chartered Surveyors or the Pyramus & Thisbe Club. Contact information is provided in the bibliography.

# 2 The Building Regulations and building control

The meaning of 'building work' is set out in Regulation 3(1) of the Building Regulations 2010 (S.I. 2010/2214) for England and Wales. The definition is a broad one and, with few exceptions, a loft conversion – and certainly one containing habitable rooms – falls within it. It should be noted that the Building Regulations are law, not guidance; powers to enforce the regulations are contained in the Building Act 1984. The penalty for non-compliance includes a fine of up to £5000 plus the requirement to pull down or alter any work that is carried out in contravention of the regulations.

## THE BUILDING ACT 1984

The Building Act 1984 is the enabling Act under which Building Regulations are made. It provides the broad legal framework for building control and gives the Secretary of State the power to make Building Regulations.

## THE BUILDING REGULATIONS

The Building Regulations – sometimes referred to as the Principal Regulations – set out procedural and functional requirements. The functional requirements are broad and include, for example, requirements relating to structural stability, fire safety and energy efficiency. Little practical detail is provided.

Amendments to the Building Regulations are frequent: Between 2000 and 2010, 19 sets of amendments were made to the Building Regulations. A consolidated version of the Building Regulations – the Building Regulations 2010 – came into force on 1 October 2010. Consolidation means the regulations are, for the time being at least, much easier to use. However, it cannot be considered to be 'light touch' legislation: there are 54 regulations and 6 schedules in the 2010 document compared to just 24 regulations and 3 schedules in the 2000 edition.

Schedule 1 of the Building Regulations 2010 sets out a number of requirements with brief explanations. With the possible exception of Part M, all are potentially relevant to loft conversions:

- Part A Structure
- Part B Fire safety
- Part C Site preparation and resistance to contaminants and moisture

*Loft Conversions*, Second Edition. John Coutts.
© 2013 John Coutts. Published 2013 by Blackwell Publishing Ltd.

- Part D Toxic substances
- Part E Resistance to the passage of sound
- Part F Ventilation
- Part G Sanitation, hot water safety and water efficiency
- Part H Drainage and waste disposal
- Part J Combustion appliances and fuel storage systems
- Part K Protection from falling, collision and impact
- Part L Conservation of fuel and power
- Part M Access to and use of buildings
- Part N Glazing – safety in relation to impact, opening and cleaning
- Part P Electrical safety

## Approved Document guidance

The Approved Documents, so called because they are approved and issued by the Secretary of State, provide practical guidance that supports Schedule 1 of the Building Regulations. There are Approved Documents to support each of the 14 parts of Schedule 1 listed above, and an additional document to support Regulation 7 – *Materials and workmanship*. In tandem with the Building Regulations they support, the Approved Documents are frequently updated.

The connection between Building Regulations and the supporting Approved Documents is not seamless. Legislation tends to run ahead of practical guidance on implementation; Building Regulations are amended far more frequently than the supporting Approved Document guidance. References to the Building Regulations in the Approved Documents – even in cases where both are published on the same day – are therefore not always consistent.

Much of the practical guidance offered in the Approved Documents is drawn from British and European Standards, codes of practice and other published technical sources. Note that supporting standards and guidance may be withdrawn or superseded within the life of an Approved Document.

It is emphasised that each of the Approved Documents is intended only to address the requirements of the regulation it supports. While each is internally consistent, it should not be assumed that it meshes seamlessly with guidance offered in other Approved Documents.

Note that some expressions used in the Approved Documents have meanings that apply only within those documents (the definition of 'habitable room' is an example – see Glossary). Even common terms should not be taken at face value. For example, in Approved Document A, the word 'wall' is used in its accepted perpendicular sense while in Approved Documents B and C, 'wall' can also mean a part of a roof pitched at more than 70° to the horizontal.

Approved Document guidance offers a number of valuable concessions without which some conversions would not be possible. These include

*Approved Document B – Fire safety (2006 edition)*
- Modified 30-minute standard of fire resistance for existing first floor (subject to conditions – see Chapter 4)

*Approved Document K – Protection from falling, collision and impact (1992)*
- Reduced headroom for stair access to lofts
- Use of alternating-tread stairs in certain cases

*Approved Document L – Conservation of fuel and power (2006 edition)*
- Insulation requirement may be reduced where it would affect floor area (but see Chapter 10)

## Compliance guides

Compliance guides are a relatively recent innovation. Like the Approved Documents, these are published by NBS on behalf of the Department for Communities and Local Government, and contain additional guidance that supplements information provided in Approved Documents.

As with the Approved Documents, by following the guidance set out in compliance guides (and any other documents referred to by an Approved Document) there is a legal presumption of compliance with the Building Regulations. Compliance guides containing information relevant to loft conversions include:

- Domestic Building Services Compliance Guide (2010)
- Domestic Ventilation Compliance Guide (2010)

## Relationship between the Building Regulations and the Approved Documents

Building Regulations and the Approved Documents are not the same thing, although they are often confused. The Approved Documents are often wrongly described as 'building regs'. In simple terms, the Building Regulations are law and the Approved Documents are guidance. The use of Approved Document guidance is not mandatory provided that work complies with the Building Regulations.

In practice, however, the relationship between the Building Regulations and the Approved Documents is rather more complex. The Approved Documents are an important instrument for the implementation of government policy and, as such, there is an increasing emphasis on quantified standards. Some of the Approved Documents are prescriptive in character and in some cases (Approved Document L, for example) it would be misleading to suggest that only guidance is offered because specific targets are set out.

There are also important legal presumptions associated with the use of Approved Document guidance. Section 7 of the Building Act 1984 states that if in any proceedings, civil or criminal, it is alleged that a person has contravened a provision of the Building Regulations, then failure to comply with an Approved Document may be relied upon as tending to establish liability. Equally, proof of compliance with the Approved Documents will tend towards negative liability.

## BUILDING CONTROL

There is an obligation to use a building control service when carrying out building work, although certain types of 'building work' are covered under competent person self-certification schemes, which allow self-certification of compliance with Building Regulations.

The purpose of building control is to provide an independent check that building work complies with the Building Regulations (but see also *Statutory notifications*, p. 25). Checks on compliance are chiefly achieved through site inspections but, depending on the procedure chosen, plans may also be inspected before construction commences. Building control services are offered by local authorities and also by private approved inspectors. In either case, there is a charge for the service. It is emphasised that building control and planning are separate, independently administered functions. Even though a loft conversion may not require planning approval, it will be subject to Building Regulations if habitable rooms are to be built.

## Local authority building control

Local authority building control is provided by unitary authorities, district councils and London boroughs in England, and by county and county borough councils in Wales. It is invoked either through a full plans application or under a building notice.

Because loft conversions are seldom straightforward, some local authorities specifically recommend that they be carried out under the full plans procedure.

Note that the Building (Local Authority Charges) Regulations 2010 gives local authorities greater flexibility to set prices for chargeable functions. Under the previous regulations, building notice and full plans charges had been equal. The 2010 regulations allow differential charging. One of the reasons for this is that the workload associated with managing a building notice can be greater than that required for a full plans application. In some cases, full plans fees are less than building notice charges, but this is not the case universally.

## Full plans

Drawings, including specifications, structural engineers' calculations and other relevant construction details, are deposited with the local authority before work commences. The application is checked and the local authority must issue a decision within five weeks. This may be extended to a maximum of two months by agreement. Plans may be:

- Approved
- Approved subject to conditions
- Rejected

Where the local authority is not satisfied with the application, it may ask for amendments to be made or additional details to be provided. Alternatively, it may issue conditional approval. This will either specify modifications which must be made to the plans, or it will specify further plans that must be deposited. The local authority may only apply conditions if the person carrying out the work has either requested it to do so or has consented to it doing so. A request or consent must be made in writing. If plans are rejected, the reasons are stated in the notice.

If plans comply with the Building Regulations, the local authority will issue a notice stating that the plans have been approved. A full plans approval notice is valid for three years from the date of deposit of the plans. If building work does not start within this time, the local authority may issue a notice stating that the approval is of no effect.

In all cases, it is necessary to inform building control of commencement at least two working days before work starts. It is possible to start work before a decision on the application is made, provided that a commencement notice is issued, but the protections offered by full plans do not become active unless a notice of approval is issued.

The advantage with full plans approval is that it reduces uncertainty about the technical specification of the work before it starts. Another potential benefit is that under full plans, it is possible to seek a determination from the Secretary of State if the plans, or any part of them, are rejected. The disadvantage with the full plans procedure is that the approval process can take some time.

## Full plans application

The full plans application form is intended to elicit only basic information. It is, therefore, incumbent upon the person carrying out the work to provide as much detail as possible in order to indicate compliance with *all* the relevant Building Regulations. This information is usually provided in drawings, plans, specification sheets and structural engineering calculations.

While there is no such thing as a 'model' full plans application, attention to the following points (some of which are set out in Regulation 14 of the Building Regulations) is recommended.

*Block (location) plan*: usually to a scale of 1:1250, showing the size and position of the building and boundaries of the site, as well as neighbouring roads. This is important in relation to fire safety because it allows an assessment to be made about potential external access for rescue purposes and possible unprotected areas in relation to relevant boundaries.

*Plans and sections*: these should be to a scale of 1:100 or better (preferably 1:50) and should cover not only the conversion but also every other floor in the building. Both existing and proposed elements of the building should be included. This is of great importance in a loft conversion where the layout and structure of even remote elements of the building can have a considerable impact on the design and indeed the feasibility of the conversion, particularly with regard to fire safety and structural stability.

Note that on a purely practical level, there are risks associated with unintentional re-scaling caused when electronically generated drawings are manipulated and printed. See Chapter 1, *Drawings general*, on ways to avoid this.

*Structural engineering*: calculations justifying the dimensions, weight, position, fixings and bearings of any beam either timber or steel. In a typical loft conversion, calculations would generally be required for the following elements:

- Floor beams
- Roof beams
- Trimmers (floor and roof)
- Header beams over windows
- Spine wall, if proposed for new load-bearing use

*Electrical safety*: some local authorities require information on whether or not the person carrying electrical work is a member of one of the prescribed self-certification schemes.

### *Common problems with full plans applications*

Approval of plans may be withheld for a number of reasons, either because detail is inadequate or missing, or because the proposal appears to contravene the regulations. Some of the more common reasons are listed below:

- Dimensions of structural members such as floor and roof beams with supporting engineering calculations absent or inadequate (Part A)
- Fire safety details including failure to provide an adequately protected stairway (Part B)
- Staircase and landing details with particular attention to headroom (Part K)
- Insufficient thermal insulation (Part L)
- Details of electrical works inadequate or statements absent (Part P)

Of the points listed above, the provision of an adequately protected stairway and the availability of headroom on staircases/landings are potentially the most difficult to remedy.

### *Pre-application prior to full plans application*

A number of local authorities now operate a plan-checking service which may be used *before* a full plans application is made. A charge is made for this service. Comments made or advice given by local authorities under these circumstances do not constitute formal approval.

## Building notice

The alternative to making a full plans application is to conduct work under a building notice submitted to the local authority. Note that in the case of loft conversions, local authorities prefer the full plans approach to be used. Less detail is required in a building notice and the information needed is set out in Regulation 13 of the Building Regulations 2010. It includes:

- A description of the building work and the location of the building
- A 1:1250 scale plan indicating the size, position and curtilage of the building and its relationship with adjoining boundaries
- Width and position of any street on or within the curtilage of the building
- Number of storeys
- Drainage provisions

Additional details may be requested. Work can commence within two clear working days of the notice being submitted. A building notice ceases to have effect three years after it is given to the local authority, unless building work commences before that time has elapsed. Like building work carried out under the full plans procedure, construction conducted under a building notice is subject to site inspections.

The disadvantage with the building notice procedure from a builder's perspective is that it is not possible to ascertain in advance whether the proposed work complies with the Building Regulations because detailed plans are not submitted. It should also be noted that the local authority is not required to issue a completion certificate under the building notice procedure.

### *Obligations under building notice procedure*

Using a building notice does not remove the requirement to provide whatever information is necessary to demonstrate compliance with the Building Regulations (Regulation 13(3)). Many local authorities insist on plans and sections of the whole building in order to demonstrate compliance with fire safety and headroom requirements. In addition, structural engineering calculations are nearly always required for loft conversions. The guidance provided by one London local authority is as follows:

> Although detailed plans are not necessary as part of the submission, structural calculations and details may be required to justify certain elements of work. For example, structural calculations will always be required for a loft conversion.

## Notification and inspection of work

### *Statutory notifications*

Whether the work is conducted under a local authority building notice or a full plans application, the Building Regulations impose an obligation on the person carrying out the work to provide notification before and after certain stages. The primary obligation is for the person carrying out the work to notify, rather than for the local authority to inspect. Notification periods are set out in Regulation 16 and they are:

- Before commencement of work (2 days)
- Before covering up any excavation for a foundation (1 day)
- Before covering up any foundation (1 day)
- Before covering up any damp proof course (1 day)
- Before covering up any concrete or material laid over a site (1 day)
- Before covering up any drain or sewer to which the regulations apply (1 day)
- After laying, haunching or covering any drain or sewer with respect to which Part H of Schedule 1 imposes a requirement (notice to be given no more than 5 days after completion of work)
- Where a building, or part of one, is to be occupied before completion (notice to be given at least 5 days before occupation)
- After completion of building work (notice to be given no more than 5 days after completion)

### *Site inspections*

As noted above, the Building Regulations 2010 sets out requirements for notification rather than inspection. In practice, of course, site inspections are carried out to ensure compliance.

While there is no formal list of inspections (other than those implied in *Statutory notifications* above), the following is provided as a 'model' inspection schedule in the case of a loft conversion. In the interests of efficiency, some of the following inspections are, of course, combined:

- Commencement
- Steel beams in loft area and any raising of chimney stack

- Floor, wall and roof timbers; location of escape windows
- Insulation to roof and walls including ventilation of voids; first fix electrical
- Drainage of sanitary fixtures; extractor fans and other ventilation arrangements
- Fire doors, smoke detectors and fire alarms; other fire resistance measures; stairs, headroom, handrails and room ventilation
- Completion

Loft conversions are generally built at considerable speed and it is important to ensure that work is not covered up before inspection. The order of inspections must reflect the sequence in which work would be concealed. Note that under Regulations 16(6), 45 and 46, the local authority has extensive powers to investigate and test the building fabric, workmanship and materials used.

### *Certificate of completion of work*

Under a full plans application, the local authority may issue a certificate of completion of work provided that it is satisfied that the work complies with the Building Regulations and that it was requested to do so when the plans were initially submitted. In the case of a building notice, the local authority is not required to issue a completion certificate, although most will. A completion certificate does not provide conclusive evidence of compliance. However, it is a legal document and its significance should not be underestimated: note that a completion certificate is normally required by a purchaser's solicitor when a house is sold.

## Resolving Building Regulations disputes

Most minor disagreements between a person carrying out building work and the building control service over whether or not plans comply with the requirements of the Building Regulations are settled by discussion and, where necessary, alterations to plans or works. However, where it is not possible to reach an accommodation, the Building Act 1984 provides a number of procedures for resolving disputes. These include the following.

### *The determination procedure (under full plans only)*

Where the building control body (i.e. a local authority or an approved inspector) believes that plans do not comply with one or more of the Building Regulations, but the person carrying out the building work contends that they do, it is possible, in England, to apply for a determination to the Secretary of State for Communities and Local Government. In Wales, determination applications are handled by the Welsh Assembly Government.

In general, the application must relate to work that has not commenced. An application can be made at any time after the deposit of plans, although clearly it is prudent to wait for either an informal indication by the local authority that the plans are not acceptable, or a formal notice of rejection.

Under the determinations procedure, which varies depending on which building control system is used (i.e. local authority or approved inspector), the arguments of both parties are taken into consideration and a decision (determination) is made as to whether the applicant's proposals comply with the Building Regulations. For example, an applicant may consider that the provision of a sprinkler system in lieu of an enclosed staircase would comply with Requirement B1, *Means of warning and escape*. The local authority's case might be that a fire-resisting enclosure is required. The Secretary of State considers both sets of arguments and makes a determination.

Appendix B contains some examples. Note that the determination procedure is not available for building work carried out under a building notice – it only applies where a full plans application has been made.

Between 1998 and 2011, the Secretary of State dealt with 79 determinations. Nearly half of these concerned loft conversions. In all but four cases, the Secretary of State determined that the proposals did *not* comply with the Building Regulations.

Note that in the first instance, the use of voluntary alternative dispute resolution procedures is now encouraged, rather than an immediate application for a determination.

### Relaxation and dispensation (full plans or building notice)

The Building Act 1984 and Regulation 11 of the Building Regulations 2010 give local authorities the power to relax or dispense with requirements contained in the Building Regulations. It is, therefore, possible to apply to the local authority to seek a *relaxation* or a *dispensation* of one or more requirements. An application to relax or dispense with a requirement may be made under either full plans or building notice, and may be sought at any stage before, during or after the building work. If an approved private inspector is being used, an application for relaxation or dispensation is made to the local authority.

There is a critical difference between the determination procedure and relaxation/ dispensation. From the applicant's point of view, the determination procedure is based on the premise that a proposal *does* meet the requirements of the Building Regulations. But under relaxation/dispensation (in which the decision is made by the local authority rather than central government), the applicant must recognise that a proposal *does not* comply with a particular requirement, or part of one.

When seeking a *relaxation*, it is necessary to state the reasons why a specific part of a requirement (usually in Schedule 1 of the Building Regulations) is believed to be too onerous; a *dispensation* is sought where a Building Regulations requirement is considered inappropriate or unreasonable.

A formal notification of the local authority's decision should first be obtained, in writing, before any further action is taken. The local authority is required to make a decision on relaxation and dispensation applications within two months. This period can be extended by agreement.

If a local authority refuses to relax or dispense with a requirement, there is a right of appeal under section 39 of the Building Act 1984 against that decision (see *Appeals*).

Many of the requirements of Schedule 1 of the Building Regulations are life safety matters and it should be noted that the consensus amongst most building control professionals is that it is not possible to relax any of the requirements, only to dispense with parts of them – and this is rare.

It is also important to differentiate between the law (the Building Regulations) and the supporting guidance (the Approved Documents) in this context. While it may not be possible to relax a requirement of the Building Regulations, there is no formal requirement to relax adherence to the Approved Document guidance, provided the applicant is able to demonstrate to the building control body that the proposed work meets the functional requirements of the regulations in some other way. This might include, for example, proposing a fire engineered solution to meet the requirement of Part B1 of the Building Regulations, rather than following the guidance in Approved Document B.

### Appeals

Appeals against a refusal to relax or dispense with one or more requirements of the Building Regulations are submitted, in England, to the Secretary of State for Communities and Local Government. In Wales, appeals are made to the Welsh Assembly Government.

An appeal must be made within one month of the date of being notified of a refusal. A written formal notification of refusal, including reasons, should be obtained from the local authority before an appeal to the Secretary of State can be submitted.

The basis of any such appeal must be an acceptance that proposed (or completed) building work does *not* comply with regulations – what is sought is a relaxation or dispensation of the regulations. This is still widely misunderstood, and an analysis of the arguments advanced in appeal reports indicates that some of those making appeals are under the impression that the work they are proposing complies in some way.

Between 1998 and 2011, more than 90 appeals were considered. All but five were dismissed. Almost one third of the appeals concerned loft conversions. In several cases, the appeals were dismissed on the grounds that the proposals had the potential to comply, and therefore it was not necessary to grant either a relaxation or a dispensation. See Appendix B for decisions concerned with loft conversions.

## Electronic building control applications

This facility, known as Submit-a-Plan, allows applicants or agents to submit building control applications and supporting documentation, including drawings, online. Where the system is fully supported by a local authority, the following electronic applications may be made:

- Full plans submission
- Building notice
- Regularisation certificate
- Replacement doors and windows

Submissions made in this way have the same status as those made under the traditional hard-copy system. An electronic application has the obvious advantage of near-instantaneous delivery and it reduces the costs associated with producing and managing paper documentation.

Submit-a-Plan can also be used to generate off-line paper documentation, such as application forms, for postal submission to local authority building control.

## Approved inspector building control

Using a building control service is a legal requirement, but there is no obligation to use the service provided by the local authority; a private approved inspector may be used instead.

Supervision of building work by individuals and organisations other than local authorities was made possible by the Building Act 1984. Currently, more than 70 companies and individuals offer an independent building control service.

Approved inspectors are bound by procedures set out in the Building (Approved Inspectors etc.) Regulations 2010. However, the functional regulations and requirements administered (i.e. those set out in the Building Regulations 2010) are of course the same for both local authority and private approved inspectors. Lists of approved inspectors are maintained by the Construction Industry Council (CIC) and by the Association of Corporate Approved Inspectors (ACAI).

# 3 External forms

This chapter considers different types of conversions and the factors that influence their choice. A number of basic forms are described. These are simplified somewhat for the sake of clarity and it is emphasised that hybrid forms incorporating several different elements are common. Drainage is considered in this chapter because it has a significant but sometimes overlooked impact on the appearance of the conversion: many otherwise reasonable designs are spoiled through lack of attention to the external arrangement of discharge stacks, branches, vents and rainwater goods.

*Note that a proposal to alter the roof may require planning permission. All habitable conversions must comply with the Building Regulations. In the majority of cases it is necessary also to invoke the Party Wall etc. Act 1996 before undertaking work. Planning and other regulatory requirements concerning loft conversions are covered in detail in Chapters 1 and 2.*

## PRIMARY INFLUENCES ON FORM

A loft conversion is never a blank slate. Physical factors (such as the form of the existing roof) and regulatory considerations (planning law) play a major role in determining design. In some respects, these constraints simplify the design process but only in the sense that they provide absolute limitations.

### Planning considerations

The influence that planning legislation has on the form of the conversion depends on the building itself and also the area it is in. Further details are included in Chapter 1. A synopsis is provided below:

- *Permitted development*: restrictions are based on height, volume and the conversion's relationship with the eaves and the highway.
- *Conversions requiring planning permission*: these will need to conform to local design guidance for planning permission to be granted in most cases.
- *Conservation areas/Article IV*: conversions altering roof shape require planning permission. Provision of roof windows may also require planning permission. Local guidance on acceptable dormer forms may be available.

*Loft Conversions*, Second Edition. John Coutts.
© 2013 John Coutts. Published 2013 by Blackwell Publishing Ltd.

## Pitch, plan and headroom

The relationship between the pitch of the existing roof slopes and the plan of the building determines the availability of headroom in the conversion. Where headroom (other than at the ridge) is restricted, adopting a box dormer design may be the only viable approach.

Note that simply being able to stand up in the loft space before conversion is not a reliable guide to final headroom. A new structural floor is likely to reduce headroom by between 150 to 250 mm, while an insulated flat roof is likely to be significantly lower than the base of the existing ridge board.

## Stair access

Even where the roof space provides sufficient headroom, it is still essential to make a separate assessment of headroom on the new stairway. Consideration should be given to the position of roof slopes or downward projecting elements (such as purlins and floor beams) that are likely to foul the new access stair.

## Shallow-pitched roofs

In roofs with a relatively shallow pitch, a rear box dormer is the most effective means of providing daylight and headroom. Note that the effectiveness of traditional small dormer forms in admitting daylight is considerably reduced in low-pitched roofs: as pitch decreases, cheek length increases to form a dormer 'tunnel' (cheeks may be glazed but see *Design considerations*). Traditional dormers work better and are generally more pleasing to the eye where the main roof slope has a pitch greater than 40°.

## Existing roof type

The structural form and arrangement of the existing roof will play a significant part in determining what type of conversion is technically and economically feasible.

### *Roof form – gable*

Roofs with gable walls, including the pitched roofs of terraced dwellings with compartment walls that extend to ridge height, are generally the easiest to convert (Fig. 3.1a). Apart from providing a greater internal volume for a given floor area (compared to hipped roofs), gable walls are often capable of providing ready-made support for new elements of structure such as roof and floor beams.

### *Roof form – hip*

Hipped roofs are more troublesome to convert than gabled roofs (Fig. 3.1b). In an unmodified state, the hip or hips restrict the useable floor area of the conversion. Dwellings with hipped roofs usually require substantial modifications if useable clearances are to be achieved. A complete replacement of the roof hip and construction of a new gable (hip-to-gable) will be needed in most cases.

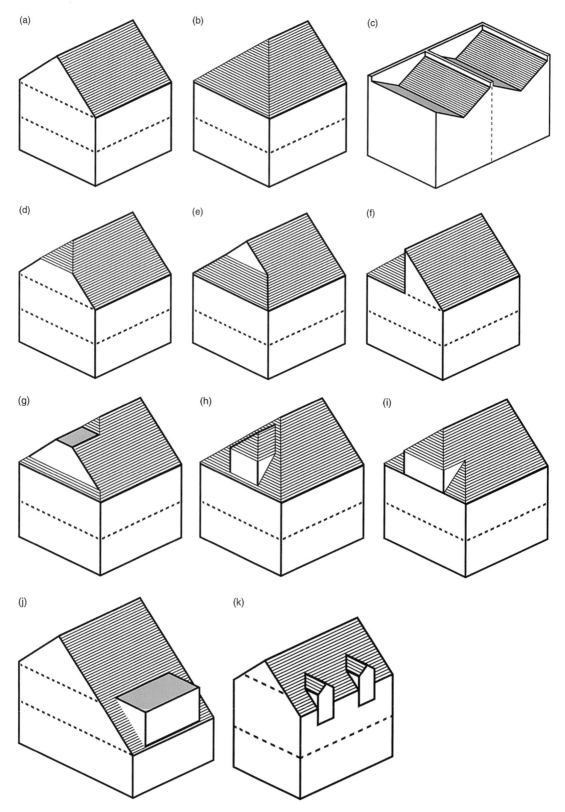

**Fig. 3.1** Roof and gable forms. (a) Gable, (b) Hip, (c) Butterfly roof, (d) Half-hipped or 'barn end', (e) Hip with gablet, (f) Split gable, (g) Compound roof, (h) Side dormer (subordinate), (i) Side dormer (plane of wall), (j) Lean-to or outshut (k) Half dormer.

### Roof form – butterfly or London roof

Because there is not a great deal of 'loft' to convert in these shallow V-shaped roofs (see Glossary for technical description), it is generally necessary remove the roof slopes entirely to accommodate a new floor and roof structure (Fig. 3.1c). Typically, a mansard-type solution is adopted on the highway-facing side. Planning permission is required because the height of the roof is increased substantially in most cases.

## CONVERSION FORMS

A number of common basic configurations are described below. Some of these are traditional 'whole roof' forms (such as the mansard) while others are roof adaptations (such as the hip-to-gable conversion). For the sake of clarity, traditional subordinate dormer forms are considered separately in the next section.

## Roof space only conversion

This describes a conversion that exploits the void bounded by the existing roof slopes (Fig. 3.2a). It is sometimes also described as a rooflight, attic or VELUX conversion. Appropriate structural modifications are carried out in order to create habitable accommodation, but these do not result in alterations to the external shape of the roof. Daylighting is provided by roof windows or rooflights set into the roof slope rather than dormer projections. This approach is generally only suitable where the roof space is of an adequate size, with headroom available over a reasonable floor area.

## Box dormer conversion

The box dormer is the most popular form of loft conversion in conventional pitched roof dwellings (Fig. 3.2b). It is capable of providing a relatively large amount of useable space quickly and at a low cost. To qualify as permitted development, a box dormer is built on the opposite side of the building to the highway.

In a typical configuration, the conversion's rear elevation (the dormer face) rises vertically near the eaves and in many cases may occupy a large proportion of the width of the roof. The triangular sides of the dormer (the cheeks) generally rake into the remnants of the existing roof slope. Both face and cheek walls are generally constructed from timber studwork with tile, slate or sometimes timber cladding. Occasionally, sheet metal coverings are used. A flat roof spans from below the ridge of the existing roof to the dormer face. The flat roof is generally laid to fall in the direction of the rearward face to simplify rainwater drainage and collection via guttering.

The provision of a flat roof provides an extended area of full-height ceiling within the conversion. This is particularly advantageous in conversions of small to medium-sized dwellings with a relatively shallow roof pitch, where useable headroom, and therefore floor space, would otherwise be restricted to the area immediately beneath the ridge.

Note that the box dormer is more or less the only practical option in small to medium-sized dwellings: most of the traditional pitched dormer forms described later in this

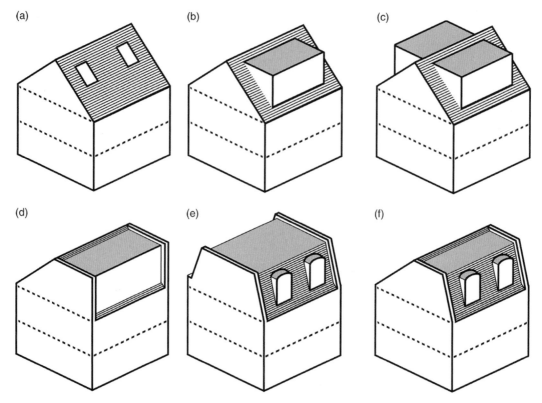

**Fig. 3.2** Loft forms. (a) Roof space only, (b) Rear box dormer, (c) Front and rear box dormers (front dormers no longer accepted under permitted development), (d) Full-width conversion, (e) Front and rear mansard conversion, (f) Rear mansard conversion.

chapter are dependent on the principal roof having a relatively high ridge if they are (a) to generate useable floor space and (b) to make any sort of aesthetic contribution.

## Front box dormer conversion

Front box dormers attracted a considerable degree of adverse comment during the 1980s and, largely on aesthetic grounds, highway-facing dormers were removed from permitted development in 1988 (Fig. 3.2c). Any projection beyond the plane of a highway-facing roof slope now requires planning permission and this is generally denied by local planning authorities where front box dormers are proposed.

## Hip-to-gable conversion

The hip-to-gable conversion, in which an existing pitched roof hip is replaced by a new vertical gable wall, usually at the side of the building, is sometimes carried out as a conversion in its own right in order to enlarge an existing roof space (Figs 3.3a and b). Generally, though, hip-to-gable conversions are carried out in order not only to increase the volume available, but also to provide support for new structural members such as roof beams, and sometimes new purlins, as part of a larger box dormer conversion. In addition, such a

**Fig. 3.3**   Hip treatments. (a) Terrace end – unmodified hipped roof, (b) Terrace end – hip-to-gable conversion, (c) Terrace end – side dormer conversion.

conversion may be necessary to provide headroom for a new staircase. A new gable end may be formed in two ways:

- *Masonry* – the gable end is built up in brick to match the existing. Alternatively, blockwork may be used and given a render finish.
- *Studwork* – the gable end is built up in timber studwork. The stud structure may be battened out for tile hanging. Alternatively, expanded metal lathing may be fixed and a render finish applied. Appropriately designed, a timber studwork gable is capable of providing support for beams and other structural elements.

Note that a number of variant hip roof treatments are possible. These include complex compound forms (Fig. 3.1d–i). See also *Side dormer conversion*.

## Side dormer conversion

The side dormer may either be set within the roof slope or built up on the head of the gable wall (Figs 3.1 h and i and 3.3c). Like the hip-to-gable conversion described above, it replaces part of an existing roof hip. Side dormers are sometimes adopted where a full hip-to-gable conversion is not acceptable for planning reasons and they are often built to provide headroom clearance for a new stairway (note, though, that hip-to-gable conversions are now considered to be permitted development in England in most cases – see Chapter 1). Where built off the gable wall, a side dormer may also serve to accommodate and conceal the ends of floor beams that might otherwise project beyond the plane of a roof slope.

## Full-width dormer with masonry flanks

This form is similar to the rear box dormer and offers similar advantages in terms of extending useable floor space (Fig. 3.2d). The principal difference is that the dormer occupies the full width of the roof, with flank walls providing a direct vertical continuation of the gable or party walls beneath: the conversion is thus bounded by masonry flank walls rather than tile hung cheeks. As with the box dormer, the dormer face is vertical and generally clad with tile or slate.

This approach is best suited to terraced dwellings that are separated by party wall parapets: compartment walls of this sort project above the roof slope and walls may be raised without the necessity of disturbing the roof covering of the adjoining property.

Raising party wall parapets to form new flank gable walls maximises space within the conversion and introduces a degree of vertical discipline to the building's rear elevation. However, because the face of the dormer and its supporting masonry flanks are closely aligned with the existing rear wall, such a conversion will tend to overwhelm the building beneath. Note that this form of conversion is sometimes, incorrectly, described as a mansard conversion (see below).

## Mansard conversion

The defining quality of the traditional mansard roof is its broken slope, with a long, steep lower section and a shorter and flatter upper slope. The mansard's double slope rises from the eaves and is applied to each elevation of the building. When applied only to opposing elevations, it is sometimes described as a gambrel or gabled mansard.

The term 'mansard' is now applied rather vaguely to any roof with a steeply pitched principal slope. In the case of a mansard loft conversion, a steeply pitched lower slope is applied to the rear elevation and sometimes to the front as well. Where it is applied to an urban front elevation (Fig. 3.2e), it may rise from behind a parapet wall (see *Butterfly roof conversion*). The modern form usually has a near-flat upper slope and is sometimes called a half mansard.

Because the face is inclined, the mansard is effective in reducing the apparent mass of a converted roof space. Properly detailed, the mansard is perhaps the most pleasing of the wide loft forms.

Mansard conversions are sometimes applied to the rear elevation only, with gables or compartment walls extended upwards to form flank gables (Fig. 3.2f). As with full-width dormers with masonry flanks (above), this approach is most easily adopted in terraced dwellings, where the existing party wall parapets project above the roof slope. For the sake of appearance, the new flank gable walls are generally raked back to match the pitch angle of the mansard. The mansard slope, which is usually inset from the masonry flanks, is generally clad with slate or tile. Lead, zinc or copper cladding may also be used.

Daylighting within the conversion is provided either by vertical windows (which by virtue of the sloping mansard face will be miniature dormers in their own right), by roof windows set in the same plane as the mansard slope or by recessed windows. In contemporary practice, the upper roof slope is generally configured as a conventional flat roof laid to fall towards the face; guttering and fascia are, of course, not provided at the junction between the slopes. Care should be taken in detailing the junction. A lead flashing fixed beneath the upper slope roof covering, dressed over and clipped to the principal slope, is, for example, preferable to a welted felt downstand.

Note that there are a number of technical requirements associated with roof pitch: part of a roof pitched at an angle of more than 70° to which persons have access is defined as an external *wall* rather than a roof for the purposes of Approved Document B *Fire safety*.

A similar pitch-based distinction is recognised in Approved Document C *Site preparation and resistance to contaminants and moisture*. Note that a roof with a pitch of more than 70° (such as that forming the lower slope of a mansard roof) should be insulated as if it were a wall.

### *Butterfly roof conversion*

In a number of inner-London boroughs, adopting a mansard-style slope is sometimes a planning requirement on highway-facing (i.e. front) elevations when dwellings with shallow V-shaped butterfly roofs are converted (Fig. 3.4). Note that an alteration of this sort is generally considered to constitute a roof extension rather than a loft conversion for the purposes of planning and building control because all or most of the roof is removed.

When this approach is adopted, party wall parapets on each side are built up to form flank walls, while to the front elevation a mansard slope is created behind the parapet wall, with rainwater collection provided by a box gutter. To the rear, the V-shaped gap between each half-gable is filled in with brick or studwork and raised to full height. A flat roof is provided, generally with party wall parapets configured to project above it.

Front elevation (proposed)    Rear elevation (proposed)    Typical cross section

Front elevation (existing)    Rear elevation (existing)

**Fig. 3.4**   Butterfly roof conversion. In this example, both the front and rear mansard slopes are pitched at 70°. Courtesy South London Lofts Ltd.

In order to preserve unity, particularly in a terrace where other conversions are antici-pated or have been carried out already, local planning authorities will generally specify the angle of pitch on highway-facing slopes. Typically, a slope angle of 70° to the horizontal is required. Note that work of this sort always requires planning permission because it alters the shape, height and visibility of the roof.

## Lean-to conversion

Some dwellings have a rear single-storey ground level outshut accommodated beneath the roof slope (Fig. 3.1j). Projections of this sort were sometimes created intentionally at the time of construction, but are often the result of subsequent extension.

Exploitation of the void formed by the roof at first-floor level is potentially rather more straightforward than a conventional conversion, if only because there is generally no need to provide an additional staircase. In a contemporary context, a suitable outshut might be formed by a garage at the side of a dwelling. It is emphasised that the floor of a loft conver-sion or any other room formed over a garage must provide full (rather than modified) 30-minute fire resistance.

## Half dormer

This term is used to describe a dormer window that occupies the junction between an external wall and the roof, for example, in the top floor of a 1½ storey dwelling where upper rooms are formed partly within the walls and partly within the roof space of the house (Fig. 3.1 k). It is not a 'pure' external dormer form in a strict sense: most of the traditional forms considered presently in this chapter may be configured as half dormers. Because it rises directly from the wall below, the face of a half dormer is generally con-structed from masonry rather than studwork.

## Existing attic rooms

Slate was rapidly adopted as an urban roofing material from the 1790s. One of the conse-quences of this, initially, was lower roof pitches (often less than 30°) and therefore roof voids that were generally unusable for habitation. However, steeper pitches became fash-ionable from about 1860 onwards, and in larger dwellings this led to the creation of sub-stantial roof voids that could be exploited. Attic rooms, with rooflights and dormer windows to provide light and ventilation, are thus original features in many larger Victorian houses.

A degree of confusion surrounds the regulatory status of original habitable attic rooms in older buildings. In some cases, purpose-built attic rooms might satisfy, or perhaps even exceed, current structural guidance, but that is all: in their 'as-built' form, it is unlikely that they would meet other current standards, for example, those governing insulation. There is no requirement that they do, *unless*:

■ The purpose to which the building is put is changed (a *material change of use*, see Glossary).
■ Material alterations are undertaken. For example, the installation of a new WC/bathroom would have to conform to current standards in an 'as-built' attic. However,

there would generally not be a requirement to upgrade anything else, even though it may be prudent to do so. Regulation 3 of the Building Regulations 2010 recognises that even if elements of a building do not comply with existing regulatory requirements, any alterations undertaken should not make those elements 'more unsatisfactory'.

## Galleries and platforms

Where a roof void is open to the room or rooms beneath it, a raised floor may be constructed to exploit the space. A gallery (sometimes called a *sleeping gallery* or a *half loft*) is thus created. Access is provided by stairs or, under some circumstances, by a fixed ladder with handrails. Guarding is required at the gallery edge.

Floor platforms of this sort are suited to certain types of single-storey dwellings (although not exclusively so), including barn conversions. In Wales, a sleeping gallery formed in the roof void is called a *crog loft* and it is a traditional feature in some small rural dwellings, particularly in west Wales. Although of some antiquity ('conversions' of this sort are probably as old as the pitched roof itself), the form remains popular (see also Chapter 4 *Galleries*).

## TRADITIONAL DORMER FORMS

The primary function of the dormer window is the provision of light and ventilation in the roof space (Fig. 3.5). Traditional dormer forms are generally subordinate to the overall roof structure. Correctly proportioned and appropriately positioned, they may enhance the appearance of the building, but note that the pitched forms are less practical in smaller dwellings where the principal roof has a low ridge.

Traditionally, the relatively low mass of small dormer windows meant that they often required support only from rafters and trimming members. However, it is emphasised that the depth of structure needed to accommodate insulation material and the requirement to provide thermally efficient double or triple-glazed windows means that modern versions of such traditional forms result in structures of greater mass. Additional means of support for the dormer (including support by the floor) must be taken into account.

Careful attention should be given to the choice of cladding and roofing material: large slates and tiles, unless they are a feature of the principal roof, often appear clumsy on small dormers. This applies in particular to the more complex hipped and eyebrow roof forms. Other cladding and roofing materials that may be considered include lead, zinc, copper, timber and glass reinforced plastic (GRP). Equally, expanded metal lathing may be fixed to the vertical studwork, and render applied.

Dormer cheeks may be glazed, rather than clad, in order to improve daylighting. Where this is proposed, the questions of overlooking and the practical difficulties of cleaning the external face of the glazing panels, which are generally fixed, must be addressed.

Traditional dormer forms are generally described by roof rather than window shape. Note that the roof of the dormer, particularly if it is gabled, is sometimes described as the *dormer head*. The following are 'pure' rather than hybrid forms.

**Fig. 3.5**   Traditional roof adaptation. The modest proportions of these half dormers and the relatively steep roof pitch contribute to a pleasing whole.

## Gabled dormer

This is perhaps the most widespread of the traditional dormer forms, with a symmetrical pitched roof, vertical studwork cheeks and a casement or sash window to the face (Fig. 3.6a). Valley detailing is required at the junction between dormer and principal roof slopes. Depending on headroom, the ceiling within the dormer itself may be fixed directly to the rafters (i.e. following the roof line) or to horizontal ceiling joists. Gabled dormers are sometimes described as *pitched*, *cottage* or *bonnet* dormers.

## Hipped dormer

As above, but with a hip end rather than a gable (Fig. 3.6b). This form works best where the principal roof itself is hipped. The term *piended* (pronounced peended) is widely used in Scotland to describe a hipped roof. See also *Canted bay*.

## Flat dormer (small)

Structurally, this is the most straightforward of the dormer forms because there are no pitched roof components and therefore valley boards/valley rafters are not required (Fig. 3.6c). The flat roof may be finished with a covering of lead, copper or zinc; in the case of lead, roll detailing may be a significant feature. A built-up felt covering may also be used. In order to eliminate the need for guttering, which might appear clumsy on a small dormer, the flat roof may be configured with a reverse fall to the principal roof slope.

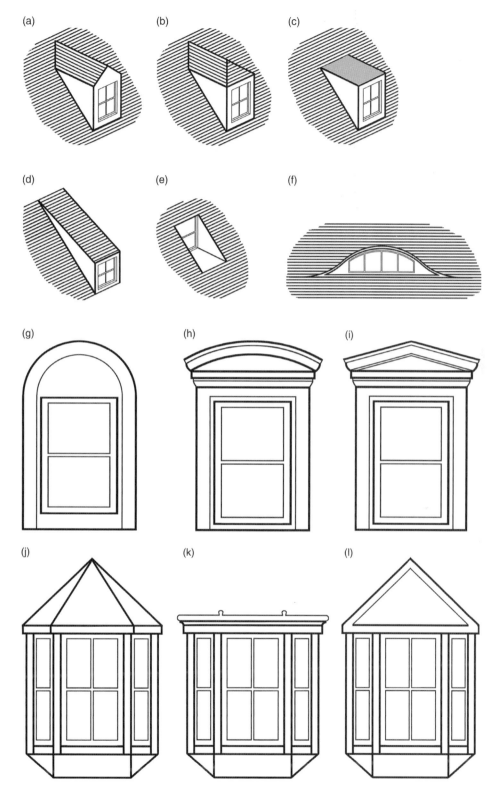

**Fig. 3.6** Traditional dormer forms. (a) Gabled dormer, (b) Hipped dormer, (c) Flat roof dormer, (d) Cat slide dormer, (e) Recessed dormer, (f) Eyebrow dormer, (g) Arched dormer, (h) Segmental dormer, (i) Pedimented dormer, (j) Canted bay dormer – hipped, (k) Canted bay dormer – flat roof, (l) Canted bay dormer – gabled.

## Cat slide dormer

This form is most effective on steeply pitched roofs (Fig. 3.6d). The cat slide dormer is a monopitched outshut of the main roof, with acute triangular cheeks, that falls to the face of the dormer at a shallower angle. Because the pitch of the dormer slope may be relatively shallow, it is important to ensure that the roofing material is capable of providing weather and uplift resistance. The cat slide is also sometimes described as an *eased-out*, *wedged* or *swept* dormer.

## Recessed dormer

The recessed dormer is sometimes adopted where planning restrictions prevent projections from the roof slope or perhaps where a small balcony is to be introduced (Fig. 3.6e). As its name suggests, the vertical window is encompassed *within* the envelope of the roof. For practical reasons, including daylighting, rainwater drainage and the impact on available floor area, this solution is best suited to roofs with a relatively steep pitch. It may be used to accommodate windows within a mansard slope. It is sometimes called an *inset* dormer. Guarding is required where it forms part of a balcony (see Chapter 8, *Juliet balconies and balustrades* and Fig. 8.20b).

## Eyebrow dormer

The eyebrow dormer has no sides, the roofing material rising and falling over the roof opening as an unbroken undulation (Fig. 3.6f). This form allows for the creation of a broad but relatively shallow rectangular window opening at the face. In carpentry terms, it is the most complex of all the dormer forms. When designing an eyebrow window, care should be taken to ensure that the roofing material is capable of conforming both to the slope and to the curvature of the projection.

## Arched dormer

This has a true semicircular roof profile: the commensurate circle would have a diameter to match the width of the dormer (Fig. 3.6 g). For practical reasons, these generally require metal cladding and, for the sake of appearance, this is often continuous across cheeks and roof. It is sometimes described as a *barrel* dormer.

## Segmental dormer

Segmental profiles are generally gently curved, with an arc struck from a point below the springing line (Fig. 3.6 h). The roof is generally finished with lead, copper or zinc. It represents a classical approach and is generally associated with front elevations. It is sometimes called a *bow* dormer.

## Pedimented dormer

The pedimented dormer is similar in some respects to the gabled dormer, at least in the sense that it is triangular in profile (Fig. 3.6i). However, like the segmental dormer described above, it is a classical feature and somewhat formal. The roof pitch of the pedimented dormer is generally shallow, in the Grecian style, and the roof is generally finished with a metal covering rather than slate or tile. It is often associated with steeply pitched roofs including mansards.

## Canted bay dormer

This is sometimes described as a *polygonal* or *bay dormer*. The front of the dormer is three sided, with a central window flanked by glazed areas that are angled back. The hipped (piended) form (Fig. 3.6j) is widespread in Scotland. Figs 3.6k and l illustrate flat and gabled versions. The gabled form tends to top heaviness. Canted bays provide good light admittance relative to dormer length.

## DESIGN CONSIDERATIONS

In most cases, the primary intention of the building owner is to create the greatest volume of habitable space at the least possible cost. An unfortunate consequence of this is that many conversions fail from an aesthetic point of view.

Many loft conversions fail to 'work' because they bear little relation to the buildings beneath them. In order to tie old and new together, the conversion should at least echo elements of the host building, and an appreciation of the existing dwelling is therefore essential. Note that the Town and Country Planning (General Permitted Development) (Amendment) (No. 2) (England) Order 2008 requires that materials used in exterior work in the enlargement of a roof must be of 'similar appearance' to those used in the existing dwelling (see Chapter 1).

Attention is drawn to the following points.

## Fenestration

New windows should respect the basic proportions of openings in the existing building. Equally, the proportions of the glazed units and thickness of glazing bars *within* individual windows should be considered in relation to those existing in lower floor windows.

The arrangement of windows should also be considered. In many cases, vertically aligning new windows with existing windows is appropriate. Where this is not possible, creating a number of evenly spaced openings is generally better than an asymmetrical window arrangement, particularly if this bears little relationship to the sequence of openings lower down the building.

In order to reduce the somewhat unrelieved appearance of the dormer face, windows may be set back into reveals, but the depth of the dormer skeleton must be configured to allow this (e.g. the use of 150 rather than 100 mm studwork). Windows should not be in the same plane as the wall unless other openings in the building are configured in this way.

## Roof detail

Both flat and pitched roofs are considered more pleasing to the eye when they oversail the walls beneath them to some extent. In the case of a typical box dormer (where a flat roof projection of about 100 mm provides visual separation between wall and roof), this is generally practical only to the rear elevation and not the cheeks.

Fascia boards should be relatively shallow and be dark in colour; white fascia boards simply draw attention to the gutter. Guttering should be returned against the dormer cheeks at each end if possible.

Box dormers tend to make single-storey buildings appear top heavy. Inclining the face of the dormer in a conversion, or setting it well back from the eaves, serves to reduce its impact to some extent.

## Vertical cladding and roofing materials

The selection of cladding material should reflect the existing roof material, but note that large format slates and tiles appear cumbersome when used to cover small areas. For practical purposes, large tiles, particularly profiled tiles, are not suited to complex detail formation, such as hips, on small dormer roofs. Plain tiles (265 × 165 mm) are the most widely available small-format covering and are often used for cladding conversions (Fig. 3.7).

Lead flashings and soakers should be treated with patination oil at the time of fixing to reduce lead carbonate staining on the cladding below them.

The pitch of new roof elements, such as gabled dormers, need not match the pitch of the principal roof exactly, unless an element of the principal roof (such as a cross gable) presents itself in the same plane as the dormer. Hipped dormers reduce the apparent mass

**Fig. 3.7**  Contemporary roof adaptation. Modern box dormer with plain tile cladding and new windows to harmonise with those on the floor below.

of the dormer structure and help it to merge with the roof slope. Note that a roof pitch of 45° is considered unattractive.

New brickwork for built-up gables should match the existing masonry. Brick type, mortar composition and bonding must be considered.

Glazed cheeks allow for increased light levels within the conversion, and fixed side glazing of this sort is sometimes used in small dormers. When evaluating this approach, the following points should be considered: fire resistance – relationship between any glazed cheek and a boundary; thermal insulation – the effects of increasing the glazed area; cleaning – providing safe access to allow cleaning of glazed side panels; and planning-potential overlooking of neighbouring dwellings.

## Chimney positions

In many cases, it is necessary to incorporate a chimney stack within the design of the conversion. Where the chimney is at the ridge of the roof, this is generally a straightforward matter. However, where the stack rises from the eaves on a rear elevation, it will be necessary either to remove it (if it is no longer in use) or build around it. Where the stack is retained and remains in use, it may be necessary to increase its height to conform to guidance set out in Approved Document J *Combustion appliances and fuel storage systems*. The dormer structure may be built to encompass the stack. Note, however, it is generally not permissible to fix structural elements to chimney stacks or breasts (see also Fig. 7.14).

## Drainage

The provision of bathroom and WC facilities must be carefully considered because the drainage of these will have a significant impact on both the internal arrangement and the external appearance of the conversion. In practical terms, the relatively broad vertical façades of most box dormers generally make the fixing of external drainage a relatively straightforward matter. By contrast, accommodating branch, discharge and vent pipes on or near the relatively narrow face of a traditional gabled dormer can only be achieved with difficulty and is likely to detract from any aesthetic gains.

Constructing the dormer face in the same plane as the existing rear wall may simplify the process of connecting the new system to an existing discharge stack. It also eliminates the often ramshackle arrangements that result when rainwater pipes and soil/vent pipes have to be offset to cross the boundary between the old and new parts of the building.

Note, however, that building up the dormer face in the same plane as the wall beneath it will tend to create an overwhelming structure. Current permitted development rules indicate that the conversion should be set back from the eaves (see Chapter 1). Consideration should be given to altering the position of the discharge stack relative to the conversion at a point lower down the building, rather than at loft level, to simplify connection.

Within the conversion, sanitary pipework should generally be kept above floor level in order to avoid conflict between pipes and elements of structure at or below floor level. It is, therefore, good practice to finalise the position of showers, basins, baths and WCs at an early stage in the design process.

Where possible, sanitary appliances should be positioned as near as possible to the external discharge stack to limit the length of branch pipes and thus ensure that pipework

**Table 3.1**   Unvented branch connections – appliance to wall/floor junction.

| Appliance | Typical branch pipe diameter (mm) | Minimum slope (mm/m) | Appliance to stack (max) (m) | Height of branch pipe above floor level at max length (approximate) (mm) |
|---|---|---|---|---|
| WC | 100 | 18 | 6 | +32 |
| Bath/shower | 50 | 18–90 | 4 | −22 |
| Bath/shower | 40 | 18–90 | 3 | −4 |
| Hand basin | 50 | 18–90 | 4 | +378 |
| Hand basin | 40 | 18–90 | 3 | +396 |
| Hand basin | 32 | 20 (at 1.75 m) | 1.75 | +415 |

Longer runs may be acceptable if ventilating pipes or air admittance valves are provided.

is able to penetrate the external wall above the level of the finished floor or other obstructions such as floor beams. The right-hand column of Table 3.1 indicates the relative height of the underside of a branch pipe to the floor at its maximum length. Cutting through structural beams to accommodate pipework is generally not feasible. Where floor joists are underslung (see Fig. 7.1), the supporting beam will be above relative floor level and may foul the passage of pipes passing through the external wall.

It is sometimes not possible to achieve the required slope for pipework for appliances with low-level drainage outlets such as WCs, baths and shower trays, particularly over extended runs or where the path of the branch pipe is obstructed by structural steelwork (e.g. by a beam at the perimeter of the conversion). Where this is the case, sanitary fixtures may be mounted on a dais to achieve the required fall. Electric macerator units capable of pumping foul water are an acceptable alternative where conventional gravity drainage is not possible from WCs.

# 4 Fire safety

This chapter considers means of escape and other elements of fire safety in loft conversions in single-family dwellinghouses. As with other aspects of the conversion process, it is emphasised that it is not possible to adopt a one-size-fits-all approach to fire safety: each conversion must be assessed individually.

## REGULATORY FRAMEWORK

Legally binding fire safety requirements are set out in Part B of Schedule 1 to the Building Regulations 2010. There are five requirements, all of which interact to some degree:

- B1 Means of warning and escape
- B2 Internal fire spread (linings)
- B3 Internal fire spread (structure)
- B4 External fire spread
- B5 Access and facilities for the fire service

Current guidance applicable to domestic loft conversions is set out in Approved Document B (AD B) *Fire safety – Volume 1 – Dwellinghouses* (2006 edition). This guidance took effect in April 2007. Additional guidance relevant to stairways is provided in Approved Document K *Protection from falling, collision and impact* (1998 edition).

There is no obligation to follow guidance provided in any of the Approved Documents if it can be demonstrated that a proposal satisfies the relevant requirements of the Building Regulations in some other way. However, in common with earlier editions, it should be noted that AD B *Fire safety* (2006) offers considerably more practical guidance than other Approved Documents. This guidance is widely used and, as far as loft conversions are concerned, adhering to it is often found to be less time consuming than proving compliance through other methods.

## MAIN CHANGES TO APPROVED DOCUMENT B (2006)

The 2006 edition of Approved Document B *Fire safety*, which replaces the 2000 edition, incorporates a number of changes relevant to loft conversions. Key changes include:

- *Protected stairway*. Most conversions now require a fully protected stairway, with fire-resisting doors. A concession described in the 2000 edition – which allowed the

*Loft Conversions*, Second Edition. John Coutts.
© 2013 John Coutts. Published 2013 by Blackwell Publishing Ltd.

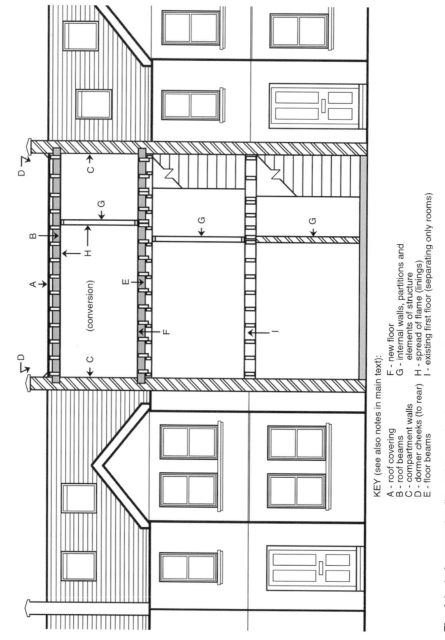

**Fig. 4.1** Loft conversion fire resistance: basic requirements.

KEY (see also notes in main text):

A - roof covering
B - roof beams
C - compartment walls
D - dormer cheeks (to rear)
E - floor beams

F - new floor
G - internal walls, partitions and elements of structure
H - spread of flame (linings)
I - existing first floor (separating only rooms)

retention of existing doors of unknown fire resistance provided they were fitted with self-closing devices – has been removed. In effect, a two-storey dwelling in which the loft is converted is now treated as if it were a new three-storey house.

- *Self-closing devices.* Other than doors between a dwellinghouse and an integral garage, fire doors need not be provided with self-closing devices.
- *Sprinkler systems.* The use of sprinkler systems in accordance with BS 9251:2005 is recognised. Sprinkler protection may be provided in lieu of an alternative escape route in dwellinghouses with a floor more than 7.5 m above ground level. Sprinklers may also be used in conjunction with other measures to compensate for the retention of an open plan layout at ground-floor level.
- *Fire alarms.* Guidance on smoke alarms has been amended and alarms should be installed in accordance with BS 5839-6:2004. All smoke alarms should have a standby power supply. Where a dwellinghouse is extended, smoke alarms should be provided in the circulation spaces.
- *Emergency egress (escape) windows.* Locks and child-resistant safety stays may be provided on escape windows. Escape windows must be capable of remaining in the open position without the need to be held open. Ladder-assisted escape via a window from a loft conversion that creates a three-storey house is no longer accepted and references to it have been withdrawn from the 2006 edition; in tandem with this, references to a maximum sill-to-eaves distance of 1700 mm for windows set in roof slopes have also been removed.
- *Galleries.* New guidance has been provided on the provision of galleries.
- *Information for occupants.* The 2006 edition places increased emphasis on providing occupants with information on operating, maintaining and using buildings in reasonable safety.

## FIRE RESISTANCE: BASIC REQUIREMENTS

Conforming to the guidance concerned with internal and external fire spread is generally a straightforward matter and compliance may be achieved through the use of adequate fire-resisting construction. Fig. 4.1 illustrates basic fire-resistance requirements in a single-family dwelling where the converted roof space creates a three-storey building.

*A – Flat roof covering.* A bitumen felt roof covering (irrespective of felt specification) is deemed to be designation AA (National class) or $B_{ROOF}$ (t4) (European class) and therefore acceptable if:

(i) the deck is a minimum of 6 mm plywood and
(ii) a finish of bitumen-bedded stone chippings covering the whole surface of the roof for a depth of at least 12.5 mm is provided.

*B – Roof beam.* Structure that supports only a roof is excluded from the definition of an element of structure in the Approved Document. It need not be protected unless it is essential for the stability of an external wall which needs to have fire resistance, or the roof is part of an escape route.

*C – Compartment walls.* Walls separating buildings must have a minimum period of fire resistance of 60 minutes.

**D – Dormer cheeks**. Depending on their relationship with the relevant boundary, 30-minute fire resistance both internal and external (see also Fig. 8.10).

**E – Floor beam**. This is an element of structure and requires 30-minute fire resistance.

**F – New floor**. Thirty-minute fire resistance.

**G – Internal walls, partitions, glazing and elements of structure**. These require 30-minute fire resistance.

**H – Spread of flame**. The use of conventional plasterboard and plaster finishes will generally satisfy the requirement to inhibit internal fire spread.

**I – Existing first floor (separating only rooms)**. Requires at least modified 30-minute fire resistance where it separates only rooms (but must provide full 30-minute resistance where it forms part of the enclosure to the circulation space between the loft conversion and the final exit – see Fig. 4.7b).

*Note that additional requirements for specific types of conversions are covered in subsequent drawings.*

## WARNING AND ESCAPE

Satisfying the requirement to provide appropriate means of escape from a loft conversion is not always a straightforward task because the relationship between the escape route(s) and rooms throughout the *entire* dwelling must be taken into account.

The guidance set out in Approved Document B is underpinned by two quite simple principles:

■ *Early warning* – an interlinked automatic fire detection and alarm system must be provided for each storey when a loft is converted.

■ *Unassisted escape* – occupants should be able to escape from the conversion to a safe place without needing outside assistance.

The risks presented by fire in a lower storey to the occupants of upper storeys increases considerably with the height of the building. Equally, increasing height has an effect on the type of escape that is possible. Thus, in broad terms, the primary escape dependency for dwellings in Approved Document B is floor height above ground level. An outline of the provisions for escape is provided below:

■ **Upper floor not more than 4.5 m above ground level**

Loft conversion typically creating a two-storey dwelling (Fig. 4.2a).
Provision of EITHER egress window(s) or external door; OR direct access to a protected stairway.

■ **Upper floor more than 4.5 m above ground level**

Loft conversion typically creating a three-storey dwelling (Fig. 4.2b).
Provision of EITHER a protected stairway OR an alternative escape route to a final exit.
*Note that egress windows in such a conversion are no longer accepted as an alternative to a fully protected stairway.*

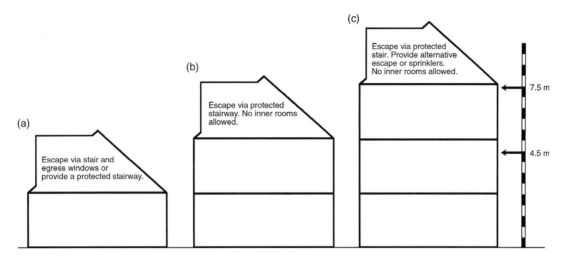

**Fig. 4.2**   Critical floor heights and associated means of escape (simplified). (a) Loft conversion creating a two-storey house, (b) Loft conversion creating a three-storey house, (c) Loft conversion creating a four-storey house.

■ *Floors at or above 7.5 m above ground level*

Loft conversion typically creating a four-storey dwelling (Fig. 4.2c).
Provision of BOTH a protected stairway and EITHER an alternative escape route OR a sprinkler system.

It should be noted that in all cases, the guidance is *less* onerous where the dwelling has more than one stairway serving the conversion, provided that the stairways provide alternative means of escape and are adequately separated from each other.

## Floor height rules

Rules for floor height measurement relative to ground level are illustrated in Appendix C to the Approved Document. The drawing in the Approved Document is reproduced here with modifications (Fig. 4.3). Assumptions are often made about the likely height of top floors based on the number of storeys in the building, but specific site factors (e.g. sloping ground or high ceilings) can have a significant impact on this. Note that the height measurement is referenced to the side of the building where the ground level is lowest.

## Storey and floor numbering rules

Ordinal storey designations ('first storey', 'second storey', etc.) should be avoided because they are ambiguous. These terms are no longer used in the Approved Document. Cardinal storey numbering ('one storey', 'two storey', etc.) is used in the Approved Document to describe entire buildings (e.g. 'a two-storey house'). Ordinal floor numbering conventions are widely understood ('ground floor', 'first floor', etc.) and these are used in the Approved Document.

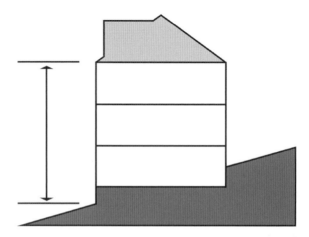

**Fig. 4.3**   Height of top storey.

## FIRE SAFETY: COMMON CONFIGURATIONS – FLOOR NOT MORE THAN 4.5 m ABOVE LOWEST GROUND LEVEL

Generally, this is a one-storey dwelling (e.g. a bungalow) made into a two-storey dwelling. Approved Document B states that the guidance for a typical one or two-storey dwelling is 'limited to the provision of smoke alarms and to the provision of openable windows for emergency egress'.

## Means of warning

A fire detection and fire alarm system must be installed. This should include at least one smoke alarm on each storey, situated in circulation spaces. In addition, a kitchen area not separated from the stairway or circulation space by a door must be equipped with a heat detector or heat alarm. The entire system must be interlinked and mains operated with a standby power supply. See also 'Fire detection' in *Elements and terminology*.

## Means of escape

Section 2 of the Approved Document outlines two sets of provisions for escape from a dwellinghouse where the upper floor is *not more* than 4.5 m above ground level. The guidance applies to all habitable rooms, except kitchens, in a dwellinghouse served by only one stair.

### Approach using emergency egress windows

All habitable rooms in the upper storey are to be provided with a window (or external door) which complies with the general provisions for emergency egress windows and doors.

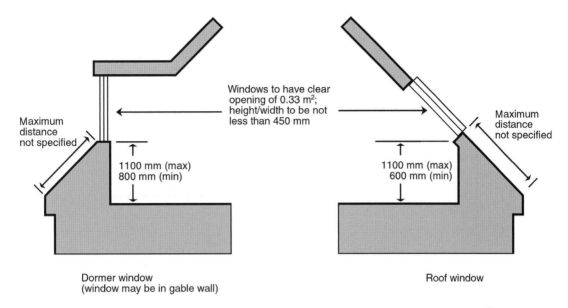

**Fig. 4.4** Emergency egress windows.
Note: emergency egress windows are only suitable for floors less than 4.5 m above ground level.

### Approach using a protected stairway

All habitable rooms in the upper storey to be provided with direct access to a protected stairway (see *Protected stairway*).

Specification of suitable windows and doors is described under 'emergency egress' in *Elements and terminology* at the end of this chapter. Note that any window of appropriate dimensions and position may be an emergency egress window (Fig. 4.4).

■ A single egress window may serve two rooms provided *both* rooms have their own access to the stairs. A communicating door between the rooms must be provided so that it is possible to gain access to the window without passing through the stair enclosure.

Where an emergency egress window is to be a roof window, the principle of self-rescue rather than ladder-assisted rescue should be held in mind. It would clearly be reasonable to consider the position of such a window in relation to the eaves. In a dwelling with a 30° roof pitch, for example, an emergency egress roof window with the bottom of the openable area set 1100 mm above floor level (the maximum permissible) would generate a sill-to-eaves travel distance of 2200 mm and the practicality of this for self-rescue would need to be considered.

In all cases, an emergency egress window or door should enable a person escaping to reach a place free from danger from fire.

### Fire resistance of new loft floor

Guidance specific to the fire resistance of the new floor in a bungalow loft conversion is not provided in AD B. It would therefore be reasonable to adopt at least the *modified 30-minute* standard of resistance specified for the upper storey of a two-storey house.

The new floor will also need to conform to guidance on sound resistance set out in Approved Document E, *Resistance to the passage of sound*. Note that the specification for achieving 40 dB airborne sound insulation might, with minor modifications, provide a somewhat higher level of fire resistance than the modified 30-minute standard recommended to satisfy fire safety requirements (see Fig. 7.17 for construction details).

## ONE FLOOR MORE THAN 4.5 m ABOVE GROUND LEVEL

Typically, this is where a two-storey dwelling is made into a three-storey dwelling. Most loft conversions fall into this category.

## Means of warning

A fire detection and fire alarm system must be installed. This should include at least one smoke alarm on each storey situated in circulation spaces, but the precise configuration will depend on the size and complexity of the dwelling, and on the means of escape provided. In all cases, the entire system must be interlinked and mains operated with a standby power supply. See also 'Fire detection' in *Elements and terminology*.

## First floor fire resistance

In the case of a two-storey dwellinghouse in which the loft is converted to create a three-storey dwelling, the existing *first* floor need not be upgraded to the full 30-minute fire-resistance standard, provided that (a) only one storey is being added, (b) the new storey contains no more than two habitable rooms and (c) the total area of the new storey does not exceed 50 m². Under these circumstances, the modified 30-minute standard is acceptable for the first floor, provided that it separates only rooms and not circulation spaces. The modified 30-minute standard is described in *Elements and terminology* below.

As noted above, the floor needs to meet the full 30-minute standard where it forms part of the enclosure to the circulation space between the loft conversion and the final exit (Fig. 4.7b).

Normally, the first floor in a three-storey dwelling is required to meet the full 30-minute standard of fire resistance. The acceptance of the modified 30-minute standard under the circumstances described above thus represents an important concession.

## New floor (conversion)

The new second floor (i.e. the conversion floor) must provide the full 30-minute standard of fire resistance.

## Escape windows

These are no longer required where the new floor is more than 4.5 m above ground level.

# Means of escape

The Approved Document outlines a number of different approaches that are capable of fulfilling the requirement to provide a means of escape from a dwelling with one floor more than 4.5 m above ground level. These are:

- Protected stairway
- Additional internal stairway
- External escape route
- Open plan escape route

*Earlier guidance, which allowed for reduced stairway protection and the provision of emergency egress windows from a new second floor, has been withdrawn.*

## *Protected stairway*

This method of escape offers a high level of passive fire protection and it is the same standard that is applied to new dwellings of three storeys (Fig. 4.5).

With this approach, a protected escape route is formed with fire-resisting construction and fire-resisting doors that leads all the way from the conversion to the final exit and to a place of safety outside the building. Note that fire-resisting doors in the enclosure need not be self-closing and that there is no requirement to provide emergency egress windows in the conversion itself.

### *Route of the protected stairway*
The protected stairway (protected at all levels) should either: (a) extend to a final exit or (b) give access to at least two escape routes at ground level, each delivering to final exits and separated from each other by fire-resisting construction and fire doors (Fig. 4.6).

Fire-resistance measures for stairways that pass through and over habitable rooms (common in dwellings with a central stair) must be carefully considered. Soffits and any other elements enclosing the protected stair (Fig. 4.7a) and any floor forming part of the enclosure (Fig. 4.7b) must provide 30-minute fire resistance.

### *Doors*
Any door forming part of the enclosure to a protected stairway in a single-family dwellinghouse should have a minimum fire resistance to the FD 20 (E20) standard. See also 'Fire doors' in *Elements and terminology*. Doors need not be self-closing.

### *Glazing*
All glazing within the protected stairway enclosure, including that in doors, walls and fanlights, needs to be fire resisting. Glazing to external windows in the protected stairway need not be fire resisting.

### *Rooflights*
Rooflights constructed from thermoplastic materials are not permissible on a protected stairway.

### *New stair*
The new storey must be served by a stair meeting the provisions of Approved Document K, *Protection from falling, collision and impact*. The new stair should be located within the protected stairway enclosure.

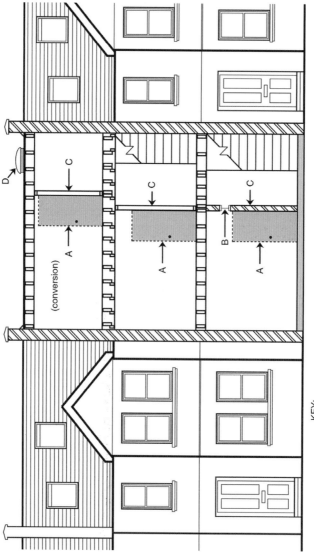

KEY:
A - FD20 (E20) fire doors (self-closers not required)
B - glazing to enclosure (e.g doors, fanlights) to be fire resisting
C - stair enclosure: 30-minute fire resistance
D - rooflights over protected stairway must be fire resisting

NOTE: cupboards opening onto the protected stairway
enclosure may require fire doors.

**Fig. 4.5**  Protected stairway enclosure.

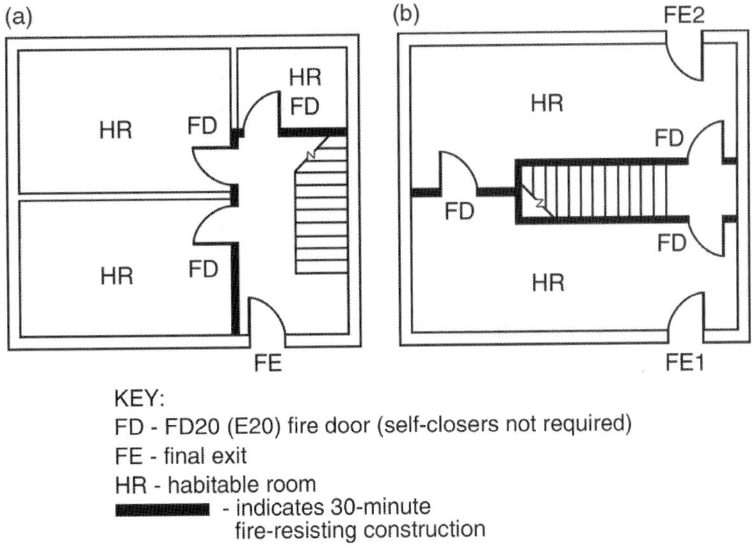

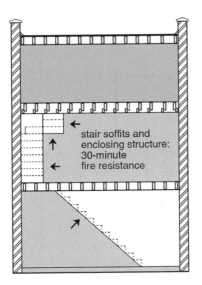

(a)
HR
FD
HR
FD
HR
FD
FE
(b)
FE2
HR
FD
FD
FD
HR
FE1
KEY:
FD - FD20 (E20) fire door (self-closers not required)
FE - final exit
HR - habitable room
- indicates 30-minute
fire-resisting construction

**Fig. 4.6** Protected stairway: enclosure and escape routes at ground level.

stair soffits and
enclosing structure:
30-minute
fire resistance

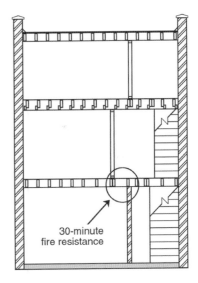

**Fig. 4.7a** Protected stairway passing through or over habitable rooms.

30-minute
fire resistance

**Fig. 4.7b** Floor forming part of the enclosure of a protected stairway.

*Air circulation systems*

See Elements and terminology.

*Cavity barriers*

Cavity barriers should be provided above the enclosures to a protected stairway in a dwellinghouse with a floor more than 4.5 m above ground level. See note in *Elements and terminology*.

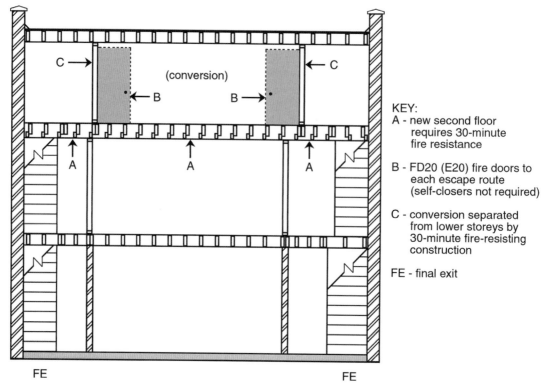

**Fig. 4.8**   Alternative escape route with additional internal stair – fire separation.

*Cupboards*

Cupboards opening onto a protected stairway should be fitted with fire doors (minimum FD20 rating) in some cases.

### Additional internal stairway

To meet the requirement using this approach, more than one stairway must be provided. The stairways should be physically separated from each other and both must provide an effective means of escape (Fig. 4.8). 'Physically separated' in this context means separated by a number of rooms, or separated by fire-resisting construction.

### External escape route

The top storey should be separated from the lower storeys by fire-resisting construction and be provided with an external escape route leading to its own final exit. An external escape route could be external escape stairs, a balcony or a flat roof. Balconies and flat roofs forming part of a means of escape must fulfil a number of criteria (see 'Balconies and flat roofs' in *Elements and terminology*).

### Open plan escape route

Under certain circumstances, an open plan layout at ground level *may* be retained if the following provisions are made:

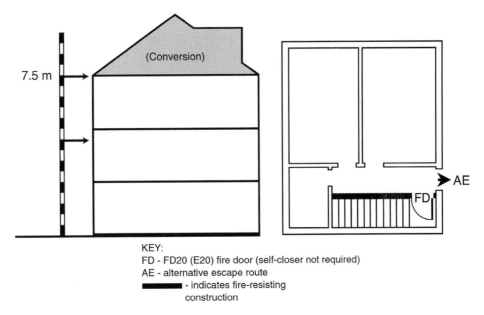

KEY:
FD - FD20 (E20) fire door (self-closer not required)
AE - alternative escape route
▬▬▬▬▬ - indicates fire-resisting
           construction

**Fig. 4.9**   Floor at 7.5 m above ground level – alternative escape.

- Sprinkler protection for the open plan area
- Cooking facilities should be separated from the open plan area by fire-resisting construction
- Fire-resisting partition and fire door (E20) to separate the ground floor from upper storeys
- Provision of an escape window at first-floor level

## MORE THAN ONE FLOOR OVER 4.5 m ABOVE GROUND LEVEL

In most cases, this is where a three-storey dwelling is made into a four-storey dwelling. The guidance in Approved Document B is that a protected stairway must be provided. In *addition* to this, either of following provisions should be made:

- Provide an alternative escape route for storeys or levels that are at or more than 7.5 m above ground level. Where access to an alternative escape route is via a landing within the protected stairway enclosure to an alternative escape route on the same storey, the protected stairway at or about 7.5 m above ground level should be separated from the lower storeys or levels by fire-resisting construction (Fig. 4.9).
- The dwelling should be fitted throughout with a sprinkler system designed and installed in accordance with BS 9251:2005 (Fig. 4.10).

## GALLERIES

For the purposes of Approved Document B, a gallery is defined as:

A raised area or platform around the sides or at the back of a room which provides extra space.

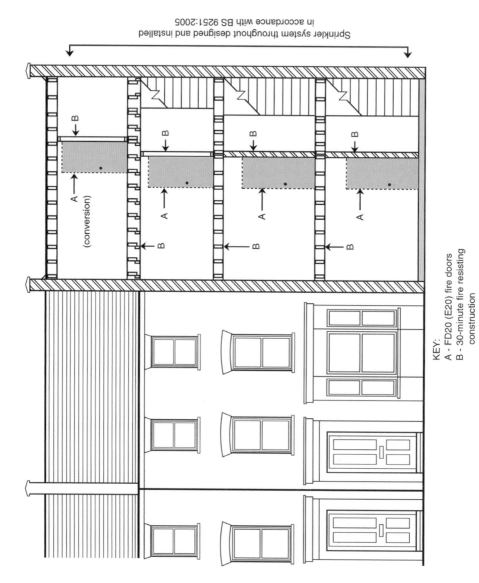

Sprinkler system throughout designed and installed in accordance with BS 9251:2005

A ← (conversion)

B

B

A

B

B

A

B

B

A

B

KEY:
A - FD20 (E20) fire doors
B - 30-minute fire resisting construction

**Fig. 4.10** Floor at 7.5 m above ground level – protected stairway and sprinklers.

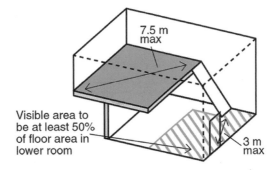

**Notes:**
1 This diagram does not apply where the gallery is
   i. provided with an alternative escape route; or
   ii. provided with an emergency egress window
      (where the gallery floor is not more than 4.5m
      above ground level).
2 Any cooking facilities within a room containing a
   gallery should either:
   i. be enclosed with fire-resisting construction; or
   ii. be remote from the stair to the gallery and
      positioned such that they do not prejudice the
      escape from the gallery.

**Fig. 4.11**   Gallery floor with no alternative exit (from Approved Document B, 2006).

Galleries in single-storey buildings are sometimes provided to exploit a roof void that is contiguous with the dwelling, or some part of it. This is probably the oldest form of 'loft' conversion. Galleries can, of course, be constructed in buildings of more than one storey.

Where the floor of the gallery is *not more than* 4.5 m above ground level, meeting the requirement to provide a means of escape is limited to the provision of an emergency egress window which must discharge to a place free from the danger of fire (see *Emergency egress windows and external doors*).

If the gallery is *more than* 4.5 m above ground level, an alternative exit must be provided (note that emergency egress windows cannot be used at heights in excess of 4.5 m).

However, in cases where the gallery floor is not provided with an alternative exit or escape window, the guidance in Approved Document B is that it should comply with the following (see also Fig. 4.11):

(a)  The gallery should overlook at least 50% of the room below.
(b)  The distance between the foot of the access stair to the gallery and the door to the room containing the gallery should not exceed 3 m.
(c)  The distance from the head of the access stair to any point on the gallery should not exceed 7.5 m.
(d)  Any cooking facilities within a room containing a gallery should either:
   (i)  be enclosed with fire-resisting construction; or
   (ii)  be remote from the stair to the gallery and positioned such that they do not prejudice the escape from the gallery.

In all cases, elements of structure supporting a gallery require fire resistance. Note also that guarding is always required at the gallery edge.

## ELEMENTS AND TERMINOLOGY

### Access room

A room through which the only escape route from an *inner room* passes.

### AFD

Automatic fire detection and alarm system.

### Air circulation systems

Air circulation systems present a risk, because under certain circumstances they may allow smoke or fire to spread into a protected stairway enclosure. For a dwellinghouse with a floor *more than* 4.5 m above ground level, the following guidance is provided in Approved Document B:

(a) Transfer grilles should not be fitted in any wall, door, floor or ceiling enclosing a protected stairway.
(b) Any duct passing through the enclosure to a protected stairway or entrance hall should be of rigid steel construction and all joints between the ductwork and the enclosure should be fire-stopped.
(c) Ventilation ducts supplying or extracting air directly to or from a protected stairway should not serve other areas as well.
(d) Any system of mechanical ventilation which recirculates air and which serves both the stairway and other areas should be designed to shut down on the detection of smoke within the system.
(e) A room thermostat for a ducted warm air heating system should be mounted in the living room, at a height between 1370 mm and 1830 mm, and its maximum setting should not exceed 27 °C.

Additional measures must be taken if ventilation ducts pass through compartment walls into another building. Guidance on this is provided Approved Document B *Fire safety – Volume 2 – Buildings other than dwellinghouses* (2006 edition).

### Alternative escape route

Escape routes sufficiently separated by either direction and space or by fire-resisting construction, to ensure that one is still available should the other be affected by fire. Note: a second stair, balcony or flat roof which enables a person to reach a place free from danger from fire is considered to be an alternative escape route for the purposes of a dwellinghouse. See also 'Balconies and flat roofs' (below).

## Automatic self-closing devices (self closers)

Other than doors between a dwellinghouse and an integral garage, fire doors no longer need to be provided with self-closing devices.

## Balconies and flat roofs

Guidance on balconies and flat roofs forming part of an escape route is set out in AD B 2.10 and 2.11. Attention is drawn to the following provisions:

- The roof should be part of the same building from which the escape is being made.
- The route across the roof should lead to a storey exit or external escape route.
- The part of the roof forming the escape route and its supporting structure, together with any opening within 3 m of the escape route, should provide 30-minute fire resistance.

Note that guarding may be needed where a balcony or flat roof is provided for escape purposes. Guidance provided in Approved Document K *Protection from falling, collision and impact* indicates that a guarding height of 1100 mm is appropriate for external balconies and roof edges in single-family dwellings.

## Cavity barriers

Approved Document B defines a cavity barrier as a construction, other than a smoke curtain, provided to close a concealed space against penetration of smoke or flame, or provided to restrict the movement of smoke or flame within such a space.

'Cavity' in this context embraces voids both large and small. For example, it can refer to the roof space over a protected stairway in a house with a floor more than 4.5 m above ground level (the 'barrier' in this case would be a fire-resisting ceiling or wall; the 'cavity' refers to the roof void over the protected stairway – although the measures outlined are not always necessary in the case of a loft conversion because most of the 'cavity' is converted into habitable space).

Cavities in walls must also be sealed, with barriers provided around door and window openings. Barriers should also be provided at the junction between an external cavity wall and a compartment wall separating buildings, and at the top of such an external cavity wall in certain cases. Equally, a barrier is required in any eaves cavity that bridges a compartment (e.g. the cavity formed by boxed eaves). Fire-resisting board or mineral wool may be used for this purpose.

## Doors – glazing in final exit

In general, it is not necessary to replace a final exit door (such as a front door facing a highway) with one of known fire resistance. However, in cases where a final exit might constitute an unprotected area (e.g. a door in the flank wall of a building) it should be considered relative to guidance provided in section 9 of Approved Document B.

## Emergency egress (escape) windows and external doors

Any window provided for emergency egress purposes (sometimes described as means of escape or MoE windows) and any external door provided for escape should comply with the following conditions:

(a) The window should have an unobstructed openable area that is at least $0.33\,m^2$ and at least 450 mm high and 450 mm wide. The route through the window may be at an angle rather than straight through. The bottom of the openable area should be not more than 1100 mm above the floor (see Fig. 4.4).

(b) The window or door should enable the person escaping to reach a place free from danger from fire. This is a matter of judgement in each case, but, in general, a courtyard or back garden from which there is no exit other than through other buildings would have to be at least as deep as the dwellinghouse is high to be acceptable.

Approved Document B also emphasises the following points:

*Guarding height*: Approved Document K *Protection from falling, collision and impact* specifies a minimum guarding height for windows of 800 mm. This is acknowledged in Approved Document B, 'except in the case of a window in a roof where the bottom of the opening may be 600 mm above the floor'.

*Locks and stays*: locks (with or without removable keys) and stays may be fitted to egress windows, subject to the stay being fitted with a release catch, which may be child resistant.

*Window operation*: windows should be designed so that they will remain in the open position without needing to be held open by a person making their escape.

Manufacturers of roof windows generally indicate which products are suitable in emergency egress applications. Top-hung rather than centre-pivot windows would normally be required in a roof slope. However, any appropriately positioned and dimensioned window is capable of serving as an emergency egress window (a dormer window, for example).

In cases where it is not possible to fix the window with its sill less than 1100 mm above floor level (perhaps because of an obstructing purlin), it is sometimes possible to provide a permanent fixed step.

## Escape route

Route forming that part of the means of escape from any point in a building to a final exit.

## Final exit

The termination of an escape route from a building giving direct access to a street, passageway, walkway or open space, and sited to ensure rapid dispersal of persons from the vicinity of a building so that they are no longer in danger from fire and/or smoke. Note: windows are not acceptable as final exits.

(a)                                        (b)

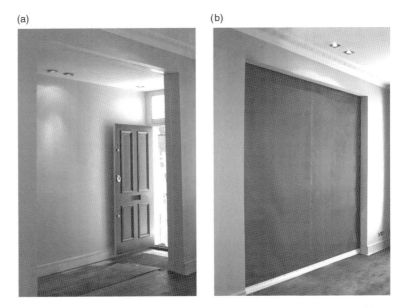

**Fig. 4.12**   Fire curtain. (a) Curtain raised, (b) Curtain closed.
Courtesy Coopers Fire Ltd.

## Fire curtains

A fire curtain comprises fire-resistant fabric fixed to a steel roller that is driven by an electric motor. The curtain is activated by a fire detection system and, guided by side channels, closes to the finished floor level. Fire curtains are used extensively in commercial applications and the technology is well established.

An appropriately specified fire curtain has the potential to provide fire separation between habitable rooms and a protected stairway in cases where there is an open plan layout at ground-floor level (Fig. 4.12). As with any fire-engineered solution, prior consultation with the building control body is essential.

## Fire detection and fire alarm systems

Fire detection and alarm systems are mandatory when new habitable rooms are provided above ground-floor level.

The following approach is appropriate for most common situations and is outlined in Approved Document B, section 1. The guidance applies in situations where ceilings are predominantly flat and horizontal:

- Provision of smoke alarms in circulation areas between sleeping spaces and places where fires are most likely to start, such as kitchens and living rooms.
- At least one smoke alarm to be provided on every storey.
- Alarms to be interlinked so that detection of smoke or heat by one unit activates the alarm signal in all of them.

Smoke alarms and detectors should be sited so that:

■ There is a smoke alarm in the circulation space within 7.5 m of the door to every habitable room.
■ Alarms should be ceiling-mounted and at least 300 mm from walls and light fittings (unless, in the case of light fittings, there is test evidence to prove that the proximity of the light fitting will not adversely affect the efficiency of the detector). Units designed for wall-mounting may also be used, provided that they are set above the level of doorways opening into the space and they are fixed in accordance with manufacturer's instructions.
■ The sensor in ceiling-mounted devices is between 25 mm and 600 mm below the ceiling (25 mm to 150 mm in the case of heat detectors or heat alarms).

Correct positioning of sensors is vital for both practical and functional reasons:

■ It should be possible to reach the smoke alarms easily and safely to carry out routine maintenance, such as testing and cleaning. For these reasons, smoke alarms should not be fixed over a stair or any other opening between floors where access is potentially hazardous.
■ Smoke alarms should not be fixed next to or directly above heaters or air-conditioning outlets. They should not be fixed in bathrooms, showers, cooking areas or garages, or any other place where steam, condensation or fumes could give false alarms.
■ Smoke alarms should not be fitted in places that get very hot (such as a boiler room) or very cold (such as an unheated porch). They should not be fixed to surfaces which are normally much warmer or colder than the rest of the space, because the temperature difference might create air currents which move smoke away from the unit.

Power for a smoke alarm system should be derived from the mains electricity supply of the dwellinghouse. The mains supply to the alarms should comprise a single independent circuit at the consumer unit, or a single regularly used local lighting circuit. There should also be a means of isolating power to the smoke alarms without isolating the lighting.

Alarms must also have a standby power supply such as a battery (rechargeable or non-rechargeable) or capacitor.

Any cable suitable for domestic wiring may be used for the power supply and interconnection to smoke alarm systems. It does not need any particular fire survival properties except in large houses (BS 5839-6: 2004 specifies fire-resisting cables for Grade A and Grade B systems). Any conductors used for interconnecting alarms (signalling) should be readily distinguishable from those supplying mains power, for example by colour coding. The electrical installation should comply with Approved Document P *Electrical safety*.

Mains-powered smoke alarms may be interconnected using radio links, provided that this does not reduce the lifetime or duration of any standby power supply below 72 hours. In this case, the smoke alarms may be connected to separate power circuits.

Detailed guidance on the design and installation of fire detection and alarm systems in dwellinghouses is provided in BS 5839-6:2004.

**Large houses:** a large dwellinghouse is one which has more than one storey, and where any storey exceeds 200 m². The following measures may be considered:

■ Large dwellings of **two** storeys (excluding basements) to have a fire detection and fire alarm system of Grade B Category LD3 as described in BS 5839-6: 2004.

- Large dwellings of **three** or more storeys (excluding basements) to have a fire detection and alarm system of Grade A Category LD2 (BS 5839-6: 2004) with detectors sited in accordance with the recommendations of BS 5839-1:2002 for a Category L2 system.

## Fire doors

When considering the fire resistance of doors, the resistance of the *entire* assembly must be taken into account including the frame and ironmongery such as hinges, handles and locks. Note that the fire resistance of the door assembly is dependent upon the use of intumescent seals. Relevant fire door specifications include:

- *FD 20 (E20)*: specified for use in forming part of the enclosure to a protected stairway in a single-family dwellinghouse. Note that although the FD20 rating still exists, relatively few FD 20 doors are now manufactured and doors meeting the FD 30 rating are used instead.
- *FD 30s (E30Sa)*: specified for use between a dwellinghouse and a garage.

In most cases, it will be necessary to replace existing doors of unknown fire resistance with certified fire doors or doorsets (Fig. 4.13). However, Approved Document B notes that if it is considered undesirable to replace existing doors (if they are of historical or architectural merit, for example) it may be possible to retain them or upgrade them to an acceptable standard.

Provided the existing doors are in reasonable condition, methods of upgrading include finishing with intumescent paint and/or the application of fire-resistant panels and stiles on the room side of the door. Hardboard or other lightweight flush doors would generally not be suitable for upgrading.

In addition to the modifications described above, building control bodies prepared to consider door upgrades will often expect compensatory mechanisms to be provided as well. These might include an enhanced fire detection and alarm system incorporating smoke detectors in all habitable rooms, and the provision of egress windows at first-floor level. In all cases, consultation with the building control body would be necessary before any work is undertaken.

## Fire stopping and the protection of openings

A fire stop is a seal provided to close an imperfection of fit or design tolerance between elements or components. It is designed to restrict the passage of fire and smoke by filling the gap.

In the case of a loft conversion, fire stopping and similar measures to seal openings in the building fabric may be required to maintain the effectiveness of fire-separating elements such as compartment walls and the enclosure to a protected stair. Similarly, other elements requiring fire resistance, such as floors, may require extra protection where the ceilings below them are penetrated by light fittings.

*Compartment walls*: fire stopping is required where timber beams, joists, purlins, rafters and battens are built in to or carried through or over a compartment wall separating adjoining dwellings

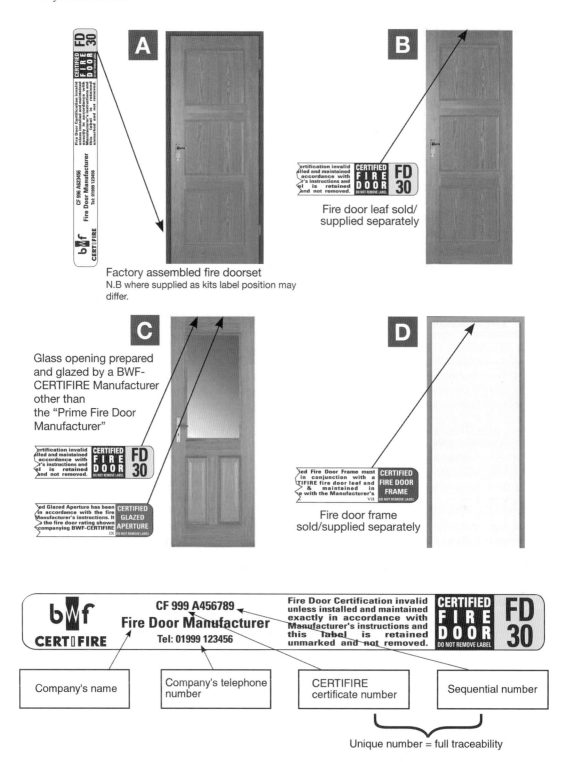

**A**

**B**

Fire door leaf sold/
supplied separately

Factory assembled fire doorset
N.B where supplied as kits label position may
differ.

**C**

Glass opening prepared
and glazed by a BWF-
CERTIFIRE Manufacturer
other than
the "Prime Fire Door
Manufacturer"

**D**

Fire door frame
sold/supplied separately

Company's name

Company's telephone
number

CERTIFIRE
certificate number

Sequential number

Unique number = full traceability

**Fig. 4.13**   Fire door certification. Courtesy British Woodworking Federation. For a colour version of this figure, please see the colour plate section.

*Protected stair enclosure*: fire stopping is required in places where pipes, cables and ducts penetrate fire-separating elements. Openings for sockets and switches, which are likely to compromise the fire resistance of stud walls, may be protected with intumescent switch and socket linings.

*Floors*: openings in ceilings, for example those created by flush-mounted downlights, must be protected to maintain floor fire resistance. Intumescent downlight covers (hoods) may be used for this purpose.

In addition to proprietary fire-stopping and sealing systems, commonly used fire-stopping materials include cement mortar, gypsum-based plaster, glass fibre, crushed rock and intumescent mastics (use depends on circumstances).

## Habitable room

A room used, or intended to be used, for dwellinghouse purposes. For the purposes of Approved Document B *Fire safety*, a habitable room includes a kitchen but not a bathroom.

## Inner room

A room from which escape is possible only by passing through another room (the access room). Inner rooms (other than a kitchen, laundry room, utility room, dressing room, bathroom, WC or shower room) are not acceptable on floors more than 4.5 m above ground level. For example, in a loft conversion that creates a three-storey dwelling, a bedroom or study whose sole means of access was via another bedroom would not be permitted. Note that the guidance on galleries differs slightly (see *Galleries*).

## Inner inner room

A room that is accessible only via an inner room.

## Loft conversion

Neither the Building Regulations nor the Approved Documents provide a formal definition of loft conversion. In broad terms, however, a loft conversion could be defined as a new storey created by conversion of an existing roof space.

A conversion that substantially *replaces* an existing roof structure, a new front and rear mansard, for example, is sometimes not considered to constitute a loft conversion for the purposes of building control. Although it refers to a specific case, attention is drawn to the following extract from an appeal dated 26 July 2001 (reference 45/3/148): 'The guidance in Approved Document B for loft conversions relates only to the conversion of an existing roof space and, as such, would not be applicable to this case as the new habitable rooms are being created by a vertical extension and the replacement of the existing roof by a mansard roof structure'.

## Modified 30-minute protection

The modified 30-minute standard for floors satisfies the test criteria for the full 30-minutes in respect of load-bearing capacity, but allows reduced performance for integrity and

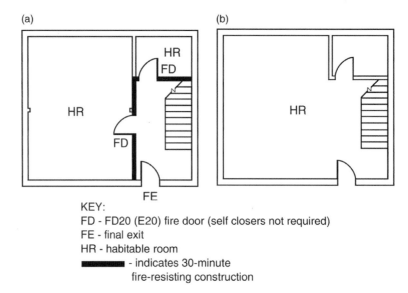

**Fig. 4.14**   Ground floor room layouts relative to stairway. (a) Through lounge: acceptable relationship with protected stairway, (b) Open plan layout – unacceptable relationship with stair.

insulation. Generally, it applies only to first-floor construction in two-storey dwellings. However, it is also permitted where a loft conversion creates a three-storey dwelling under conditions outlined earlier in this chapter. 'Modified' in this sense does not necessarily mean that anything need be done to the existing floor, provided that it is capable of meeting the standard. Note that the floor for the conversion itself (the second floor) would require full 30-minute fire resistance.

## Open plan layouts

Open plan layouts, often adopted at ground-floor level in single-family dwellings, have become increasingly popular since the early 1970s. These fall broadly into two types:

*Through lounge*. In a typical 'through lounge' or 'knock-through' configuration, the wall separating a pair of previously independent habitable rooms, perhaps a kitchen and a dining room, is removed to create a large single space (Fig. 4.14a). Provided that the new, larger, room is *not* open to the stairway (other than by a fire door), such an arrangement is less likely to prejudice the conversion of the roof space.

*Open plan*. In a full 'open plan' configuration (Fig. 4.14b), however, the walls separating habitable rooms from the stair are removed. It should be noted that fires tend to originate in habitable rooms rather than circulation areas and, where habitable rooms and circulation areas have been merged by the removal of partition walls, the risks to the occupants of the building from fire and smoke are considerably increased.

The Approved Document notes that where an open plan arrangement exists at ground level, it may be necessary to provide a new partition to enclose the escape route. In certain cases, however, a package of compensatory measures may be considered (see *Open plan escape route* p. 58, and *Sprinkler systems*).

## Passenger lifts

Where a passenger lift is provided in a dwellinghouse and it serves any floor more than 4.5 m above ground level, it should either be located in the enclosure of the *protected stairway* or be contained in a fire-resisting lift shaft.

## Sprinkler systems

Approved Document B now makes specific reference to sprinkler systems and notes that such systems can sometimes be used as a compensatory measure where the provisions of the Approved Document are varied in some way. The inclusion of sprinkler systems in AD B 2006 provides valuable clarification and removes much of the uncertainty that had sprung up around the subject in earlier years.

The Approved Document makes reference to two specific situations where sprinkler protection, in conjunction with other measures, could be applicable to a loft conversion. In both cases, sprinkler systems must be designed and installed in accordance with BS 9251:2005 *Sprinkler systems for residential and domestic occupancies. Code of practice.*

*More than one floor over 4.5 m above ground level.* Instead of providing an alternative escape route, the dwelling may be fitted throughout with a sprinkler system.

*Open plan layout at ground-floor level.* In some situations, the sprinkler protection of an open plan area at ground-floor level would allow elements of an open plan arrangement to be retained.

Automatic sprinkler systems are normally deployed as a whole-building solution. But the Approved Document notes that where sprinklers are installed as a compensatory measure to address a specific risk or hazard, it may be acceptable to protect only part of a building.

Water for an automatic sprinkler system may be provided direct from the service main. However, in areas where mains pressure and/or flow are not sufficient, a small pump and tank are installed to guarantee supply. Water is distributed to sprinkler heads via pipework which is typically concealed in ceiling voids. Each sprinkler operates independently and incorporates a heat-sensitive plug, so only the sprinkler or sprinklers in the immediate vicinity of the fire operate.

## Storey exit

A final exit, or a doorway giving direct access into a protected stairway, firefighting lobby or external escape route.

## Storey height measurement

The height of the top storey is measured from the upper floor surface of the top floor to ground level on the lowest side of the building (Fig. 4.3).

## FIRE SAFETY IN CONTEXT

Improvements in building design and the widespread adoption of smoke alarms have been instrumental in reducing the number of fire deaths in dwellings by nearly 60% since a peak in 1979. Nevertheless, there were 49 600 dwelling fires throughout the UK in 2008; these resulted in 353 deaths and more than 10 000 casualties.

Mains-powered smoke alarms are mandatory in loft conversions, but to be effective, routine testing and maintenance are essential. In dwelling fires in which a mains-powered smoke alarm was present, the alarm failed to activate in 15% of cases – 2590 fires – in 2008. Reasons for non-activation included system faults and incorrect installation (10%) and acts preventing the alarm from operating (including being turned off) in nearly a quarter of cases.

Currently, neither the Building Act 1984 nor the Building Regulations impose any obligation on householders to maintain safety features such as alarm systems.

Approved Document B (2006), however, emphasises the need to provide owners and occupiers with sufficient information to operate, maintain and use buildings in reasonable safety. This includes the provision of basic advice on the proper use and maintenance of systems provided in the building, such as emergency egress windows, fire doors, smoke alarms and sprinklers. Even simple measures, such as encouraging householders to keep fire doors closed at night, are potentially life-saving.

It should be noted that unauthorised modifications to a dwelling following conversion, such as the removal of fire-resisting doors or the opening up of a ground-floor layout that compromises the integrity of a protected stairway, constitute material alterations and would therefore render the householder liable to prosecution.

# 5 Conversion survey

As noted in Chapters 3 and 4, the form of the existing building and its roof type will have a considerable influence on the design of the conversion. The condition and structural configuration are also of critical importance. A preliminary survey should be carried out to assess:

- The suitability of the roof space for conversion
- The internal arrangement of the building
- The structural configuration of the building
- The condition of relevant building fabric
- The extent of any remedial work that is required and any other work that could profitably be carried out at the same time as the conversion

A loft conversion is sometimes not feasible for practical and economic reasons. Key considerations include:

- *Headroom*: there is no minimum height within rooms but 2.3 m is considered to be reasonable. However, minimum headroom requirements apply to stairs and landings. Lack of headroom is a potentially intractable problem: raising the ridge height of a roof is not always feasible from a planning perspective. Equally, dropping ceilings in the floor below is often not economically viable. Note that a low final ceiling height (or lower than expected) is one of the commonest causes of customer dissatisfaction when lofts are converted.
- *Open-plan layout*: where the conversion would create a floor more than 4.5 m above ground level, the stair escape route must be protected by fire-resisting construction. In a building with an open-plan layout, therefore, it would be necessary either to reinstate partitions or provide compensatory measures (see Chapter 4).
- *Roof structure*: it is not always considered to be cost-effective to adapt a trussed rafter roof, although it is usually technically feasible provided there is adequate headroom available.

## SURVEY PROCEDURE

To proceed with the design of a loft conversion, it is necessary first to produce or acquire detailed drawings for the whole of the existing building with full construction details and measurements.

*Loft Conversions*, Second Edition. John Coutts.
© 2013 John Coutts. Published 2013 by Blackwell Publishing Ltd.

The importance of examining the whole building cannot be overstated. The primary reason for this is that elements remote from the conversion itself can have an impact on feasibility. For example, the relationship between the stairway, rooms and final exits at ground-floor level will have a determining influence as far as fire safety is concerned.

In some instances, especially where the building is of recent construction, original plans may be available. It should be noted, however, that there is sometimes a considerable difference between what appears on paper and what was actually built. In addition, where original plans are to be used, any subsequent alterations to layout should be carefully noted. Where it is necessary to produce drawings from scratch, as it is in most cases, plans, elevations and sections should be at a scale of not less than 1:50.

Where drawings are to provide the basis of structural calculations, detailed measurements should be provided. These might, for example, include elements such as the span, spacing and dimensions of rafters and purlins. Comprehensive and accurate measurements provided on drawings may allow a structural engineer to perform some or all of the calculations without the need for a site visit.

The survey must take into account factors that may affect the viability of the conversion and this includes the general condition of relevant building elements. It should also identify any additional works that could be carried out at the same time as the conversion, for example, retiling a front roof slope or repointing masonry.

## OUTLINE OF SURVEY ELEMENTS

The key survey elements are listed below. A more detailed examination of these is provided in the following pages.

*Age of building*
- Date of construction (including dates of subsequent extensions)

*Headroom*
- Available headroom in roof space
- Optimum positions for new staircase and landing
- Position and form of existing staircase(s)

*External relationships*
- Position and size of adjacent buildings/structures
- Ground-level extensions (existing or proposed) to the dwelling
- Loft 'terracing'

*Internal layout*
- Position, use and dimensions of rooms throughout the dwelling
- Positions of final exits

*Roof form*
- External form and plan of roof

*Roof structure*
- System of construction (cut roof, TDA truss or trussed rafter)
- Positions, dimensions, spacing and spans of all elements of roof structure

*Roof condition*
- Evidence of fungal and insect attack
- Evidence of water penetration
- Fixity of structural elements
- Condition of roofing material, fixings and underlay
- Condition of valleys, soakers and flashings

*Walls (external and internal)*
- Load-bearing walls
- Gable walls and chimneys
- Solid and cavity walls
- Lintels

*Foundations*
- Assessment
- Geotechnical factors
- Historical factors

*Internal partition walls*
- Materials and construction method
- Load-bearing internal partition walls and their foundations
- Openings (e.g. knock-throughs) and beam provision
- Internal doors and glazing to circulation areas

*Floor and ceiling structure*
- Ceilings (materials, construction and thickness) including floorboard gaps
- Floor and ceiling joist direction, dimensions, spacing and span
- Strength of existing timber elements

*Water tanks*
- Cold water storage
- Feed and expansion

*Drainage and services*
- Position of soil and vent pipe
- Position and adequacy of existing rainwater drainage
- Position of service pipes
- Position of aerials, solar collectors and other external fittings.

*Chimneys*
- Chimney breast continuity between floors (internal)
- Position and dimensions of flues/chimney offsets in roof space (internal)
- Stack position and height relative to proposed conversion (external)

## SURVEY ELEMENTS IN DETAIL

## Age of the building

The age of the building will provide an indication of likely modes of construction (Fig. 5.1). Age is also relevant from a planning perspective. For example, under England's General Permitted Development Order (GPDO), the volume of any addition to the roof since 1948 is deducted from permitted development allowances (see also Chapter 1).

## Headroom and floor-to-ceiling height

Available headroom is generally the primary determining factor in the feasibility of a loft conversion. The ability to 'stand up' beneath the apex of the existing roof is sometimes cited as a guide to the suitability of a loft for conversion. This is not a universally reliable test because the combined thickness of a new, separately supported floor and an insulated roof structure must be subtracted from the ceiling joist to ridge measurement. A new floor with 200 mm joists may exceed 250 mm in depth when boarding and a deflection gap is taken into account. A flat, warm roof structure with 170 mm joists may exceed 360 mm at the ridge end where the firring is thickest.

There is no longer a minimum ceiling height requirement *within* rooms, but for the comfort and safety of the occupants, it is generally considered desirable to create new rooms with a minimum floor-to-ceiling height of 2.3 m. However, there is a presumption that any downward projecting elements (such as beams) within rooms present a hazard if the clearance is less than 2 m.

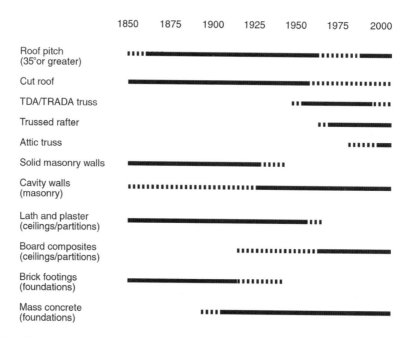

**Fig. 5.1**   Elements of domestic construction.

Approved Document K indicates explicitly that clear headroom of 2 m must be provided over the whole width of any stairway, ramp or landing. Where this cannot be achieved, the Approved Document outlines an alternative arrangement (which applies only to loft conversions) with 1.9 m to the centreline of the stairs and 1.8 m to the lowest side (Fig. 7.18b). Where the conversion provides only one habitable room, it may be possible to consider the use of an alternating-tread staircase, but only if it is not possible to accommodate a conventional stair (i.e. one with a pitch not greater than 42°).

In most cases, the new stair is positioned in the same stairwell as the existing stair. Ideally, the new stair should enter the conversion parallel to the new floor joists to minimise the need for an extensively trimmed opening (see also Figs 7.12a and 7.20).

## External relationships

A situation now often arises where a terraced house undergoing conversion may already be flanked by existing loft conversions. The survey should note the presence of adjoining conversions, and, where this is the case, a careful inspection should be made to determine the position of steel beams or other structural members.

Where steel beams have already been introduced into party/compartment walls to support an adjoining structure, it is essential to identify precisely where these have been installed. With 9" party/compartment walls, it is general practice to allow penetration and bearing of 100 mm, but there will clearly be many instances where penetration has exceeded this, or the wall is thinner than 9", and the potential for beam-to-beam conflict exists. In some cases, disturbed brickwork at ceiling joist level may provide evidence of the position of neighbouring steelwork.

## Internal layout

The relative positions of habitable rooms, the stairway and the final exit are of critical importance as far as fire safety is concerned. For the purposes of Approved Document B *Fire safety*, a kitchen is a habitable room but a bathroom is not. Note that an open-plan layout at ground-floor level, or elsewhere, will influence the viability of a loft conversion because of the need to provide a protected stairway (see Chapter 4).

## Roof form

Roofs are generally either gabled, with perpendicular walls built up at each end of the building, or hipped, in which case the roof slope is carried around the building. Both gable and hip roofs may incorporate projections such as bays with valleys at slope intersections. It is generally not practical to incorporate small projections such as bays within the envelope of the conversion, although these are often boarded-out for light storage. Note that while both hip and gable roofs may be converted, it is generally simpler to convert a gabled roof.

## Roof structure

The structural form of the existing roof will have a major bearing on the feasibility of the conversion. In undertaking the survey, the aim is to reach a full understanding of the

**Fig. 5.2**  Re-conversion. This nineteenth-century roof is undergoing its second major modification in 20 years. Note the accumulation of both structural and non-structural timber elements.

structure and function of the roof. A number of examples are illustrated in Chapter 9. Note that, in some cases, the structure may defy ready analysis (Fig. 5.2).

The internal structural arrangement of a roof takes a considerable number of forms. For the sake of convenience, roofs are often characterised as either 'cut' roofs or trussed rafter roofs. The TDA truss roof (see Fig. 9.1b) could be said to represent an evolutionary half-way stage between the two.

*Cut roofs* often lend themselves well to conversion because, generally, a relatively small number of structural elements project into the roof void. In conducting the survey, the position, dimensions and, where relevant, span and spacing of roof members should be recorded. These may include:

- Wall plates
- Purlins
- Purlin struts
- Straining pieces
- Ties
- Collars
- Binders
- Hangers
- Rafters (common, valley, hip, cripple, crown, jack, principal)
- Ceiling joists
- Ridge board

The location of purlins is often of particular importance because these will frequently obstruct new window and stairway positions.

The direction of ceiling joists must be noted. In most cases, it will be necessary to run new floor joists parallel to the existing ceiling joists, often suspended in the void between them, in order to preserve headroom.

*Trussed rafter roofs* are generally more troublesome to convert than cut roofs. With conventional trussed rafters, much of the roof void is occupied with the webs from which the roof derives a large proportion of its strength (see also Chapter 9). As well as noting truss element dimensions, method of fixing and truss spacing, the presence of bracing members should also be recorded.

Since about 2000, attic trusses (sometimes called room-in-roof or conversion trusses) have become common in new dwellings (Figs 9.1d and 9.23b). Like conventional trussed rafters, these are produced off-site by specialist manufacturers. Timber sections of considerably greater depth and breadth (e.g. $47 \times 197$ mm rather than $35 \times 97$ mm sections) are used in their production. Webbing in the conventional sense is largely absent, and attic trusses therefore have the advantage of providing a useable roof void with integral floor joists.

## Roof condition

A careful inspection should be made to identify any currently occurring or past damage to the roof's timber structure, including water penetration, fungal attack, insect infestation and rodent (e.g. squirrel) damage. Professional advice must be sought if bats are present (see *Bats*, Chapter 1).

Damaged or decayed elements may need to be replaced (Fig. 5.3). The source of any leaks, past or present, should be identified and remedied: the position of leaks is almost impossible to identify once insulation and final finishes have been applied. This particularly applies to front roof slopes, the external coverings of which are generally not modified as part of the conversion.

Evidence of fungal or insect attack will require closer scrutiny, and possibly treatment, of the entire roof structure (see also 'house longhorn beetle', Glossary).

The condition and adequacy of structural fixings should also be noted with particular reference to nailed connections between rafter feet and ceiling joists – these are an integral part of the tie between opposing roof slopes.

The type of roofing material should be noted, and careful attention paid to its condition. The state of fixings and the presence or absence of underlay/sarking should also be noted. Where it is proposed to clad the conversion in a material to match the existing roof, the feasibility of reusing slates or tiles displaced by the new dormer should be considered.

The provision and adequacy of roof ventilation should also be examined. Flush eaves (see Fig. 9.3) require careful attention because they are sometimes troublesome to ventilate, and the use of proprietary tile/slate ventilators must be considered.

Checks should also be made on the condition of valleys, soakers and flashings. Where a roof slope is to be retained, as at the front of the building, it is worth considering replacing the roof covering at the same time as the conversion. Replacing the roof covering at a later date is likely to disturb internal rafter-level plasterboard/skim finishes that are applied as part of the conversion process.

**Fig. 5.3**   Leaks and fungal attack require remediation.

Consideration should also be given to the nature of any replacement roofing material, not just on aesthetic grounds but also on loading. Replacing slates with concrete tiles, for example, will increase roof loading and may therefore have structural implications. Where re-roofing takes place, the provision of breather membrane rather than conventional underlay may be considered.

## Walls

In a terraced dwelling, the conversion loading is usually transferred via beams to party walls, or to a party wall and a flank wall if at the end of a terrace (see Fig. 6.1 for a typical beam configuration). Note that front and rear walls are often heavily compromised by existing structural openings and are generally not used to support new beams.

Any wall required to support a new load must be thoroughly assessed. Party walls have the disadvantage of being difficult to examine easily, but it is essential nonetheless that the extent of buttressing relative to length, mode of construction and thickness be determined. The same checks must be carried out on flank walls.

In some pre-twentieth-century dwellings, a party (compartment) wall separating buildings may not extend into the loft. A check should be made of the adequacy of any party/compartment wall in the roof void. This must include a check on fire-stopping between the wall and roof structure, the presence of perforations or gaps in the masonry between the dwellings, and the thickness.

External load-bearing walls should be visually inspected, noting the method of construction and general condition. Cracked masonry or evidence of bowing or buckling may require further investigation and remedial action. Similarly, mortar condition must

**Table 5.1**  Solid brick masonry walls.

| Brick fraction | Nominal thickness |
| --- | --- |
| Brick on edge | 3" |
| ½ brick | 4½" |
| 1 brick | 9" |
| 1½ bricks | 13½" |
| 2 bricks | 18" |

be noted, and walls and chimneys repointed where necessary. Note that the strength of brickwork in older buildings where lime mortar has been used (generally pre-1920) may be relatively limited. The strength of both bricks and mortar must be taken into account: an unfactored stress of $0.42 \, \text{N/mm}^2$ is sometimes assumed in these cases.

Gables should be checked for plumb, particularly where it is proposed to extend the gable to form a rear flank gable (Fig. 8.1b). Chimney stacks to be incorporated within a new wall in a hip-to-gable conversion should be similarly checked. Note that, over time, a combination of sulphate attack and weathering may cause chimney stacks, and sometimes gable walls within which chimney breasts are accommodated, to lean out of the perpendicular. This effect is caused by the expansion of mortar and it is most pronounced on the masonry face that has the greatest exposure to rain (a typical leaning chimney can be seen in Fig. 3.5).

Solid walls for dwellings were common until the 1920s and were still being used, sporadically, in the years following World War II. An indication may be provided by the bond (generally Flemish or English for solid walls). In cases where solid masonry walls are built with an outer face in stretcher bond, as is sometimes the case, tying between the inner and outer courses may be absent and the stability of the wall must be considered.

Solid brick walls are often generically described as 9" walls but note that (a) this is only a nominal designation because brick lengths varied and (b) overall wall thickness was height dependent (see Table 5.1). 'Brick on edge' and 4½" walls were generally used for internal partitions, but there are examples of party (compartment) walls of 4½" thickness. Walls of 9" thickness were widely used in external and party walls in two-storey dwellings and are commonly encountered. For wall heights of up to 40', combinations of 18", 13½" and 9" were sometimes used. Walls of 13½" thickness were also used in any dwelling where the height of one storey exceeded 10'.

Cavity walls were adopted after 1920 and were in widespread use by the outbreak of World War II, although the earliest examples appeared at the beginning of the nineteenth century. External evidence of cavity walls is generally provided by the use of stretcher bond, i.e., a running bond without headers.

Until the early 1970s, cavity walls were constructed with an outer leaf half-a-brick (about 100 mm) in thickness with a 2" (50 mm) cavity width and an inner, generally load-bearing, leaf formed from 4" (100 mm) blockwork. Later variations included the use of lightweight blockwork on the inner leaf, a 75 mm (or greater) cavity width and the use of cavity insulation, which was routinely provided from the mid-1980s. Inner and outer leaves are linked by wall ties.

In most cases, the inner leaf of a cavity wall performs the load-bearing function. In earlier masonry-constructed buildings with cavity walls (from the mid-1920s onwards), the inner leaf is generally built from brick or clinker blocks. Lightweight and, latterly, ultra-lightweight blocks have been used in the construction of the inner leaf.

Timber-framed domestic dwellings, with internal softwood wall frames that carry the load of floors and roof, have been constructed in growing numbers since the 1960s. Note that the external brick skin of a timber-framed building is generally not intended to support any of the elements of structure. However, in the case of both timber frame and cavity masonry walls, the outer leaf is designed to resist wind loads.

In all cases (both for timber frame and masonry buildings), it is necessary to check the adequacy of wall plates and, in the case of timber frame buildings, head binders and supporting studwork if it is proposed to use these to support elements of the new structure.

The position of doors, windows and other openings should be recorded as part of the survey. The general condition of wall openings should be examined, and lintels exposed, checked for adequacy and replaced where necessary. This is of importance where walls contain wide openings. Particular attention must be paid to upper-storey lintels where there may only be a few courses of brickwork between a window opening and wall plate. Note that, in many cases, bays have only limited load-bearing potential.

## Foundations

Before the widespread availability of mass concrete in the early twentieth century, the footings of most domestic buildings were relatively shallow, with stepped courses of brickwork built off a shallow concrete strip, or, in some cases, directly off the subsoil. A building's foundations and the subsoil supporting them must, therefore, be considered in relation to the proposed new loading, which is generally increased by about one third when a loft conversion is added to a two-storey dwelling.

A trial pit may be excavated to allow the footings to be checked. The form of the substructure, and particularly the projection (width) of footings, can then be examined and the bearing capacity of the subsoil assessed. Remedial measures, such as underpinning, may be required is some circumstances.

Below-ground inspections are not routinely requested by building control bodies: foundations are sometimes assumed to be adequate and this assumption is based to a great extent on the local knowledge of the building control service. In all cases, existing and proposed loads, including the extent of load dispersal at the base of walls, must be considered. In assessing the need for checks on foundations, the following points may be considered:

- Geotechnical factors
- Historical factors
- Additional factors

### Geotechnical factors

Knowledge of local subsoil, taking into account bearing capacity, potential for volume change and height of the water table. Foundation movement may be manifested by the presence of movement cracks and movement away from the perpendicular in existing external and internal walls. Conversely, it may be noted that ground conditions beneath foundations in older buildings may improve as a consequence of long-term consolidation.

### Historical factors

Including the likely or known configuration of substructure (composition, width and depth of foundations) based on the age and type of dwelling.

### Additional factors

History of local subsidence/settlement/heave particularly in clay subsoils; history of remedial works to buildings in the immediate vicinity. History of drainage repairs. Proximity of trees, noting species. Note that the localised removal of trees in clay subsoils may increase the risk of heave.

Foundation movement and damage to dwellings in the UK is most commonly caused by the expansion and contraction of clay subsoils rather than by additional loading above ground. Anecdotal evidence suggests that routine loft conversions in single and two-storey buildings do not, generally, compromise foundation stability. However, any additional loading has the potential to contribute to foundation movement, particularly if the ground is subject to volume increases or decreases caused by changes in soil moisture levels.

## Internal walls and partitions

Internal walls and partitions should be examined. There are two main reasons for doing so. One reason is to ensure (in the case of walls bounding a protected stairway) that the walls are capable of offering an appropriate degree of fire resistance. The other is to identify potential load-bearing structure: before the advent of trussed rafter roofs, a high proportion of traditional 'cut' roofs depended on the provision of an internal spine wall to transfer some roof loads to foundations, as well as to provide support for intermediate floors and, in the roof space, ceiling joists. Where suitable, spine walls are sometimes used to provide support for the floor structure of a conversion.

The only way to assess whether an internal wall is capable of supporting new or different loads is to examine the whole wall from the roof space down to the foundations. The materials used, the method of construction, the vertical continuity between floors (the walls should not be offset between floor levels) and the type of foundations must be noted. It may be necessary to expose the footings and fabric of such an internal wall to assess its load-bearing potential.

In buildings constructed after 1920, internal load-bearing walls may be formed from brick, block or timber studwork. Earlier buildings may use brick or studwork with a lath and plaster finish, and sometimes studwork with brick infill. In many cases, it will be necessary to determine the load-bearing suitability of internal walls by calculation. This will certainly be the case if a spine wall has been altered. This would apply, for example, where two rooms have been knocked together at ground-floor level and, in this case, it would also be necessary to check the adequacy of the beam provided.

**Fig. 5.4**   Service penetrations should be fire stopped.

A traditionally constructed internal stud wall built from $100 \times 50$ mm studs at 400 mm centres with 16 mm lath and plaster to both sides might provide 30-minute fire resistance without modification if the lath and plaster is in good condition. Services and fittings that penetrate such walls (e.g. pipe and cable runs, ventilation ducts and electrical sockets and switches) will all compromise fire resistance (Fig. 5.4). Fire-stopping measures, including intumescent linings for electrical fittings, may be required.

Doors between habitable rooms and the proposed protected stairway should also be examined closely. In most cases, it will be necessary to replace existing doors with new doors of known fire resistance (see Chapter 4). Conventionally glazed doors and fanlights to the enclosure are generally not acceptable because they do not offer adequate fire resistance. Conventional glazing is usually acceptable in final exit doors and external windows in a protected stairway, however (see also Chapter 4).

## Floor and ceiling structure

All existing floors and ceilings must offer an appropriate level of fire resistance (see Chapter 4). In addition, the floor structure of the existing upper floor of the building must be capable, in most cases, of providing support for a new staircase to the conversion.

The fire resistance of a floor is dependent on a number of factors. These include the composition and condition of the ceiling (Fig. 5.5), the size and spacing of joists, and the type, condition and thickness of floorboards. In addition, the presence of gaps between the floorboards of intermediate floors should be noted because these compromise fire

**Fig. 5.5**  Lath and plaster ceilings and walls: nib failure.

resistance. This applies particularly with plain edge boards, but it is also relevant with tongue and groove boards that have shrunk (see also Fig. 7.16). The position of downlight fittings and other ceiling penetrations should be carefully noted; in most cases, it will be necessary to retro-fit these with fire hoods.

## Strength of existing timber elements

The introduction of strength grades for timber is a relatively recent development. The first grading systems emerged in the late nineteenth and early twentieth centuries but these were somewhat haphazard and it was not until the immediate post-war years that formalised approaches to testing and grade marking were developed.

Existing timber elements of unknown strength can be, and sometimes are, visually graded in situ. An assessment of probable strength is based on observed characteristics, such as knots, fissures and slope of grain. Graders are trained in accordance with BS 4978:2007 *Visual strength grading of softwood. Specification.*

## Water tanks

The position, dimensions, relative height and functions of water tanks in the loft must be noted. In most cases, it will be necessary to move the tanks to make the best use of the available floor space. The position of pipes must also be noted as part of the survey.

### Cold water storage tank

This generally serves a dual function. In a conventionally plumbed dwelling, it feeds water to lavatory cisterns and cold tap outlets (but generally not the cold tap in the kitchen). Where a vented hot water cylinder is used, the cold water tank also provides a cold feed to the hot water cylinder and a termination for the hot water cylinder vent pipe.

The existing tank can be moved, if there is space to accommodate it. Potential positions include the eaves and, where headroom permits, the apex of the roof. In cases where the existing tank cannot be accommodated in this way, installation of a space-saving 'coffin' tank should be considered (see Glossary). Because the head of water (pressure) at tap outlets is a function of relative tank height in conventional systems, any new tank position that is lower than the original must be carefully considered.

An alternative, which is sometimes appropriate when the conversion is being carried out as part of a wider renovation, is to eliminate the need for a cold water tank by changing to an unvented hot water heating system (also known as a Megaflo system). Under this arrangement, the hot water system operates at mains pressure. In addition, cold taps and cisterns previously fed by gravity are switched to the mains feed.

Where an unvented hot water system is proposed, it is essential that a flow and pressure test be conducted as part of the survey to establish that the mains supply meets the unvented system's flow and pressure criteria. Most water companies aim to supply water at 1 bar, which is the minimum pressure required for most unvented systems. However, the minimum pressure water companies are obliged to maintain in the communication pipe is 0.7 bar. Note that building control must be notified where unvented systems are proposed and that such systems must be commissioned and certified by the installer.

As noted above, unvented systems eliminate the need for a cold water cistern in the loft, but space is still required within the building for the hot water storage cylinder. An alternative – where space is at a premium – is to install a combi (combination) boiler, which provides mains-pressure hot water and heating direct from the boiler with no need for a separate hot water storage cylinder.

### Central heating feed and expansion (F&E) cistern

This is an integral part of open-vented central heating systems. It keeps the system topped up and accommodates expansion of the water as it increases in temperature. In order to eliminate the need for such a tank, a closed or sealed central heating system would need to be installed.

## Drainage and services

The survey should include an accurate record of the existing arrangement of drainage and other services. The position of the soil and vent pipe is often of critical importance where the conversion is to be provided with sanitary facilities. In terms of integrating old and new, attention should also be given to the position and adequacy of the existing rainwater drainage system and guttering. Similarly, the position and orientation of aerials, solar collectors and other external fittings should be considered if there is likely to be a need to relocate them.

**Fig. 5.6** Severed chimney breast: inadequate support.

## Chimneys

Externally, the position of flues and chimneys relative to the proposed conversion and its windows should be noted and measured. The presence or otherwise of chimney cowls should also be recorded.

The position of chimney breasts within the dwelling must also be recorded in the survey. This is often a critical factor in determining the arrangement of the conversion, particularly where a steel roof beam is required to span the width of the building. It is not permissible to fix structural members, such as beams and joists, to a chimney breast.

The path of the chimney breast should be tracked through the building and, where chimney breasts have been removed, it is prudent to check that the remaining masonry is adequately supported (Fig. 5.6). In some cases, it will be necessary to retrospectively install a means of support. Corbelled brickwork or gallows brackets (see Glossary) are sometimes acceptable for this purpose, although it may be necessary to install a supporting beam spanning between walls.

# 6 Beams and primary structure

When a loft is converted, new loads must be transferred to the foundations of the existing building. These additional loads may be transmitted via beams to external walls and sometimes to internal load-bearing walls. The purpose of this chapter is to consider the characteristics, use and methods of fixing for both steel and timber beams.

*Beams and other critical structural configurations should be designed by a structural engineer or other competent person. Building control bodies will require supporting structural calculations in almost every case when a loft is converted.*

## APPROVED DOCUMENT GUIDANCE

Guidance on structural stability is provided in Approved Document A *Structure* (2004). Note that the structural design standards currently referenced in Approved Document A have been replaced by Eurocodes (BS EN 1990 to BS EN 1999) with supporting National Annexes. The following British Standards series have now been withdrawn:

- BS 5268 (structural use of timber)
- BS 5950 (structural use of steelwork in buildings)
- BS 6399 (loading for buildings)

## BEAM POSITION RELATIVE TO EXISTING STRUCTURE

Headroom is perhaps the most valuable commodity in a loft conversion. One of the most effective methods for preserving headroom is to limit the depth of the floor structure. In order to do this, it is generally necessary to limit the clear span of timber joists to about 4.5 m between supports. For a number of reasons, including the joist depth required, it is generally not feasible to exceed this by much. As most buildings are considerably deeper than 4.5 m, support for joists must be provided by beams either at one or both ends. Note that the maximum span of any floor supported by a wall is 6 m.

In loft conversions, it is practice to support the main floor beams in flank or compartment walls (i.e. generally the side walls of the building). Flank walls generally contain few openings likely to prejudice the stability of the wall (none in the case of the party walls of a mid-terrace house). This should be contrasted with the front and rear walls in most

*Loft Conversions*, Second Edition. John Coutts.
© 2013 John Coutts. Published 2013 by Blackwell Publishing Ltd.

dwellings which characteristically feature numerous openings, often with lintels of unknown strength. In addition, ceiling joists generally run parallel to flank walls: fixing floor beams at right angles to the existing ceiling joists allows new floor joists to be underslung between them.

In most cases, universal beams (UBs) or universal columns (UCs) are used in loft conversions. Timber composite and laminated beams may also be used, although their relative size means they are generally impractical for use where space is at a premium. Flitch beams, however, are used relatively widely, often as floor trimmers and in certain roof beam applications. Characteristics of both steel and timber composite beams are outlined below. In all cases, a clearance of 25 mm should be provided between floor beams and existing elements (such as lower floor ceiling joists) in order to minimise the risk of damage to retained structure caused by deflection.

## BEAM CHARACTERISTICS

Beams in loft conversions often serve more than one purpose (Figs 6.1 and 6.2). Where a floor beam is positioned towards the front of the dwelling, it may provide support for both the new floor and for the front roof slope via a purlin wall. Similarly, a roof or ridge beam may provide support for both a new flat roof and for an existing front roof slope. In many loft conversions, it will be necessary to support one end of the roof beam on a timber post; lateral restraint of the beam must therefore be considered. Subsidiary beams supported by a pair of main beams (e.g. the stair trimmer and timber post support in Fig. 6.1) are sometimes described as secondary beams.

## Common structural steel sections

Hot-rolled steel sections are widely used as beams in the process of loft conversion. Steel has the advantage of being relatively inexpensive. It is also widely used and therefore readily accepted by local authority building control when appropriate calculations are provided. The following sections are commonly used to provide support for new floor, wall and roof structures in conversions:

- Universal beams
- Universal columns
- Parallel flange channels

### *Universal beam*

UBs are produced in a range of standard sections designated by depth, width and mass. Thus, a UB designated 203 × 133 × 25 has a depth of 203 mm, a width of 133 mm and a linear mass of 25 kg/m (nominal). In all cases, the depth of a UB is greater than its width. The flanges of a UB, both internal and external, are parallel (Fig. 6.3a). 'Advance' sections manufactured by Tata Steel Europe (formerly Corus) are designated UKB, rather than UB.

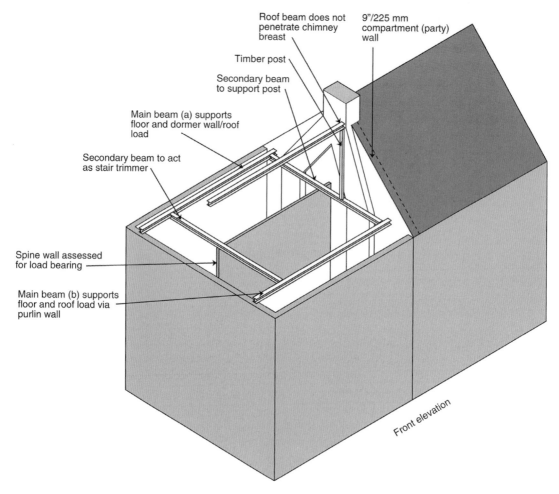

**Fig. 6.1** Box dormer: basic structural configuration (near gable omitted for clarity).

Roof beam does not penetrate chimney breast

9"/225 mm compartment (party) wall

Timber post

Secondary beam to support post

Main beam (a) supports floor and dormer wall/roof load

Secondary beam to act as stair trimmer

Spine wall assessed for load bearing

Main beam (b) supports floor and roof load via purlin wall

Front elevation

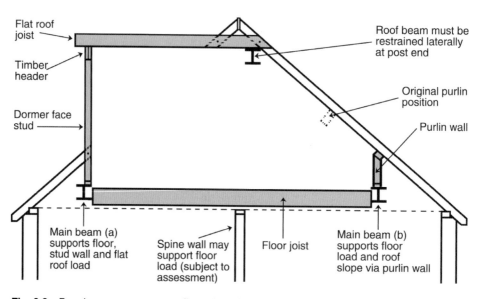

**Fig. 6.2** Box dormer: common configuration of elements.

Flat roof joist

Timber header

Dormer face stud

Roof beam must be restrained laterally at post end

Original purlin position

Purlin wall

Main beam (a) supports floor, stud wall and flat roof load

Spine wall may support floor load (subject to assessment)

Floor joist

Main beam (b) supports floor load and roof slope via purlin wall

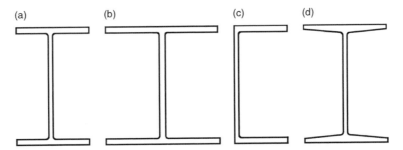

**Fig. 6.3**  Structural steel sections. (a) Universal beam (UB) practical range: 127×76 (13 kg) to 457×191 (74 kg), (b) Universal column (UC) practical range: 152×152 (23 kg) to 254×254 (89 kg), (c) Parallel flange channel (PFC) practical range: 100×50 (10 kg) to 430×100 (64 kg), (d) Rolled steel joist (RSJ) 152×127 (37 kg) and 203×152 (52 kg) only.

### Universal column

UCs are used both in vertical and horizontal configurations, designated in the same way as UBs by nominal depth, width and linear mass. Flanges are parallel inside and out. Widely used as floor beams where it is necessary to accommodate the beam within the depth of the floor structure (Fig. 6.3b). UCs used as beams deflect less than the equivalent length of UB due to their higher moment of inertia, but come with a weight penalty that makes them harder to manoeuvre in confined spaces. UC 'Advance' sections manufactured by Tata Steel Europe (formerly Corus) are designated UKC, rather than UC.

### Parallel flange channel

Parallel flange channels (PFCs) are designated by depth, width and linear mass. Typical applications include the replacement of timber purlins (Fig. 6.3c). 'Advance' sections manufactured by Tata Steel Europe (formerly Corus) are designated UKPFC, rather than PFC.

### Rolled steel joist

The term 'rolled steel joist' (RSJ) is often applied generically to steel beams. In strict terms, this is incorrect because RSJ is a designation in its own right. RSJs were largely superseded in the construction industry by the introduction of UBs and UCs in the late 1950s. RSJs have tapered internal flanges and are therefore less convenient when inserting bearers or the ends of joists into the webbing. RSJs are encountered in older buildings and are still produced in very small numbers (Fig. 6.3d).

## Engineered timber beams

The depth of section required means solid timber members are generally not practical in longer beam applications. Even in the context of joists, solid timber is generally limited to spans of between 4 and 5 m. Engineered timber beams, which may take the form of

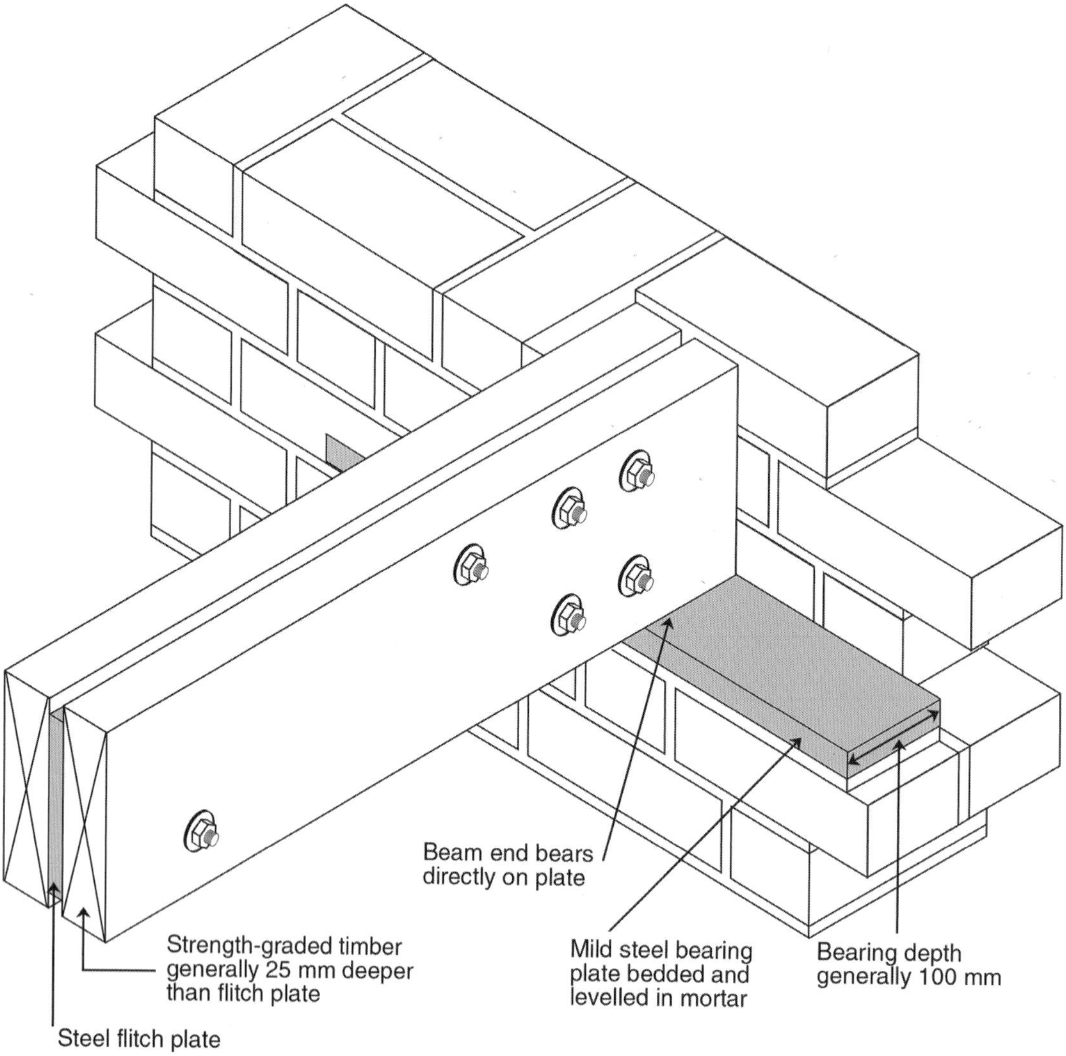

**Fig. 6.4**    Flitch beam.

laminated solid sections or fabricated composite beams, provide considerably more scope, and the characteristics of some common forms are outlined as follows:

- Flitch beams
- Laminated timber beams
- Timber composite beams

### *Flitch beams*

Flitch beams (sometimes called sandwich beams) are widely used in loft conversions (Fig. 6.4). The beam comprises a central steel plate generally between 8 and 20 mm thick

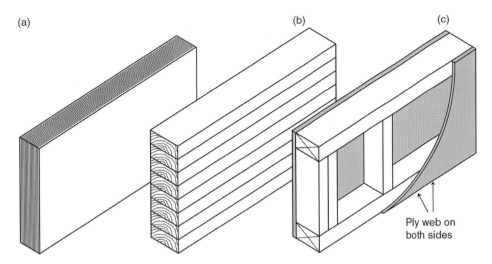

**Fig. 6.5** Engineered timber beams. (a) Laminated veneer lumber (LVL), (b) Glued laminated timber (glulam), (c) Box beam.

sandwiched between strength-graded timber sections. The flitch is the only composite beam that can be site-fabricated.

As with other beams, the use of flitch beams is subject to the provision of structural engineering calculations. However, as a rule of thumb, a flitch beam is equivalent to a solid timber section approximately 30 times the thickness of its steel plate. Bearing plates or padstones are generally required at supported ends. Flitch beams offer relatively little resistance to lateral forces; restraint is generally provided by joists positively fixed to the side of the beam. Beam components are generally fastened to each other by bolting through at intervals of not more than 500 mm.

### Laminated timber beams

A number of proprietary versions are available. These include laminated veneer lumber (LVL, Fig. 6.5a), which has relatively slender vertical laminations, and glued laminated timber (glulam, Fig. 6.5b), which is characterised by horizontal solid timber laminations with a depth of 45 mm. Note that glulam has certain aesthetic benefits. Planed and treated, it may be left exposed.

Because they are of constant section, laminated timber beams may be cut to length without fear of compromising their structural integrity. The disadvantage is that relatively deep sections are required. Fixed at low levels, laminated beams would block access to an eaves space; at higher levels, beams of this sort would clearly have an impact on available headroom.

### Timber composite beams

Timber composite beams generally incorporate softwood flanges top and bottom with deep plywood webs. Beams of this sort have the advantage of offering a high strength-to-weight ratio (Fig. 6.5c).

Given their considerable depth and the associated handling problems, composite beams are perhaps best suited to situations where a large roof is being replaced, or in new-build applications. The integrity of such beams is usually dependent on gluing, and beams are generally manufactured by specialists under controlled conditions – site fabrication is generally not possible. Equally, because they often depend on correctly spaced and prefabricated internal stiffeners, composite beams cannot generally be cut to length without appropriate – calculated – modification.

## FIRE RESISTANCE OF BEAMS

Beams supporting a new floor require a minimum 30-minute fire resistance. Note that a roof beam would generally not require 30-minute fire resistance. Approved Document B *Fire safety* excludes a structure that supports only a roof unless, for example, the roof performs the function of a floor, is a means of escape or is essential for the stability of an external wall that needs to have fire resistance.

Protection of structural elements such as floor beams may be provided in a number of ways:

- *30-minute fire-resisting ceiling below beam*: an imperforate ceiling of appropriate construction may be provided beneath the beam (see Fig. 7.3).
- *Boxing-in*: in cases where a steel beam supports a floor from beneath, it is possible to provide protection by cladding all exposed sides of the beam with an appropriate fire-resisting board (see Fig. 7.3).
- *Intumescent paint*: this is a practical consideration if a steel beam is providing support from below. It is not as effective a means of protection if joists or other elements are fixed into the webbing because the paint requires a space into which to char in order to provide protection.

Consideration should also be given to the fire resistance of the supported ends of beams. This would apply, for example, where a beam is supported by a compartment wall separating buildings (a party wall). In general, a bearing of 100 mm is provided, and in a 9" wall this would generally mean that the beam end was protected by masonry at least 100 mm thick (i.e. half a brick) on the adjoining owner's side. This is usually sufficient, provided that any gaps are thoroughly fire-stopped. Additional protection measures may be necessary where end cover is less than 100 mm, or where the depth of cover is not known.

## BEAM BEARINGS

For practical purposes, beams are generally supported at right angles to their span. However, supporting a structural steel beam or flitch beam by allowing it to bear directly onto bricks or blocks is not usually considered to be satisfactory. Concentrated loads at supported ends will create bearing stresses that may exceed permissible values and this can lead to localised failure in the masonry which may take the form of spalling or cracking. In many cases, the bearing strength of existing masonry is not known and therefore a conservative value must

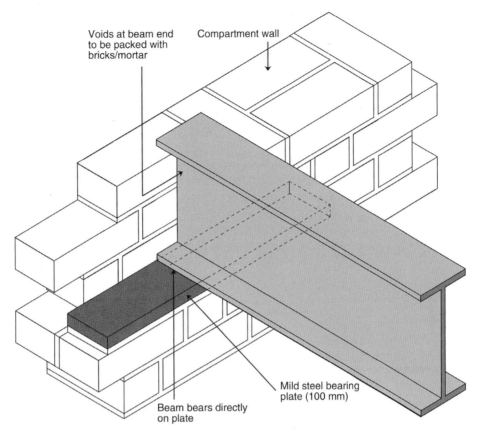

Voids at beam end
to be packed with
bricks/mortar

Compartment wall

Mild steel bearing
plate (100 mm)

Beam bears directly
on plate

**Fig. 6.6**   Beam support: mild steel bearing plate.

be adopted. As noted in Chapter 5, a value of $0.42\,N/mm^2$ is sometimes assumed. However, the overall compressive strength of masonry walls constructed from soft stock bricks bedded in lime mortar may be as low as $0.21\,N/mm^2$ (BS CP 111 part 2: 1970).

To allow a broader distribution of the load on the wall, it is normal practice for the structural engineer to specify the use of an intermediate element of known strength. This generally takes the form of a mild steel bearing plate or a padstone.

Beams should bear at full width on any padstone or bearing plate. Note that any proposal to notch or chamfer a beam end (e.g. where it must be fitted beneath a roof slope) should be supported by a structural engineer's calculations. It is not always possible to cut a beam in this way. In these cases, it is necessary to allow the beam to project beyond the roof slope, where it may be protected by lead cladding.

## Mild steel bearing plates

Where steel beams, especially floor beams, are to be installed as part of a loft conversion, mild steel bearing plates are often used in preference to padstones because there is less need to remove masonry (Fig. 6.6).

**Table 6.1**  Common sizes of concrete padstones (mm).

| |
| --- |
| 215 × 140 × 102 |
| 300 × 140 × 102 |
| 440 × 140 × 102 |
| 215 × 140 × 215 |
| 440 × 140 × 215 |
| 215 × 215 × 102 |
| 440 × 215 × 102 |

Mild steel bearing plates used in loft conversions are available in a range of standard thicknesses and are cut to length by suppliers. In most cases, calculations are based on a bearing depth (penetration) of 100 mm, which coordinates with the thickness of a single brick or blockwork leaf. Bearing plates are bedded and levelled in mortar; beam ends must bear directly on the plate.

Note that the stiffness of mild steel bearing plates is critical: plates which are not sufficiently stiff will create areas of high localised stress on brickwork beneath the beam.

Short beam sections (sometimes called cuttings) may also be used as bearing plates. In all cases, the specification of any bearing plate should be justified by calculation.

## Padstones

Padstones can be cast in situ, cast in moulds on site, or purchased from suppliers in various sizes. Table 6.1 indicates the range of padstones produced by a typical manufacturer.

Pre-cast padstones are produced under controlled conditions using concrete that is compacted by vibration to increase its density before being allowed to cure. This has the advantage of producing a component of known compressive strength – typically in excess of 50 N/mm$^2$ 28 days after manufacture. Factory-produced padstones are generally produced in sizes that coordinate with standard bricks and blocks. They generally do not contain reinforcing bars.

Padstones are fixed with mortar in recesses created in the masonry (Fig. 6.7). It should be stressed that steel beams must bear directly on the padstone with no intermediate use of mortar.

Padstones may also be cast in situ. Attention should also be given to the amount of time needed for the concrete to cure before any beam is fixed and loaded. To limit the risk of shear failure within the concrete, the depth of the padstone should exceed the offset (Fig. 6.8).

Pre-cast concrete lintels of appropriate dimensions and compressive strength are sometimes used as an alternative to padstones. However, the presence of steel-reinforcing bars in concrete lintels should be considered in relation to any subsequent requirement to drill and fix to them.

In order to accommodate a padstone, a considerable depth of brickwork must be removed. Where floor beams are to be installed close to existing ceiling joists, it will be difficult to get access to the masonry to achieve this, and there is a risk of damaging walls and finishes in the rooms below.

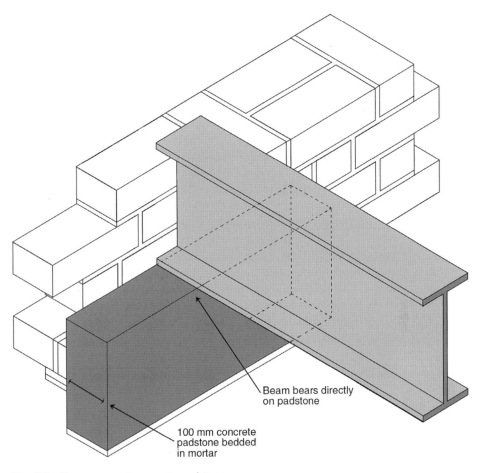

**Fig. 6.7**   Beam support: concrete padstone.

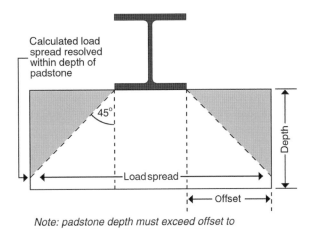

*Note: padstone depth must exceed offset to limit risk of shear failure if cast on site without reinforcement*

**Fig. 6.8**   Site-cast concrete padstone.

## BEAM PENETRATION

As noted above, beams must be adequately supported by padstones or bearing plates. However, consideration must be given to the extent of beam penetration into walls. Typically, beams for loft conversions are designed with a notional end bearing of 100 mm. This reflects the fact that the inner (load-bearing) leaf of most cavity walls is 100 mm. Equally, in a solid 9" wall, removing brickwork to the depth of half a brick would yield a similar depth of bearing. In most cases, the specification of a 100 mm bearing means that the supported end of the steel does not extend beyond the notional centreline of a 9" party wall (see also Chapter 1, *Party Wall Act* and Chapter 5, *External relationships*).

It is generally not possible to use a chimney breast to support a beam (guidance on fixing near flues is provided in Approved Document J *Combustion appliances and fuel storage systems*). This often leads to difficulties where it is necessary to install a beam at ridge height. A commonly adopted approach in these situations is to support the chimney end of the ridge beam on a timber post. The upper part of the support is generally bolted to the flange of the ridge beam while the lower part of the post is supported either by a suitable internal load-bearing wall, a spine wall for example, or by a secondary steel beam running parallel to the chimney breast (Fig. 6.1). Note that the post-supported end of the beam must be restrained to prevent lateral movement.

## BEAM SPLICES

Handling long and heavy structural steel elements in the confines of a loft space presents a number of challenges. The principal difficulty is that a wall-supported beam must be 200 mm longer than the gap it is designed to span. In semi-detached and end-of-terrace houses, it may be possible to create an opening though a gable wall or a roof hip to facilitate final positioning. In terraced houses, however, this is usually not possible.

To simplify handling and positioning in restricted spaces, beams may be cut into two or more sections with spliced connections that allow the beam to be re-assembled in situ. There are a number of methods for splicing beams in these situations. Note that some of the following approaches are not suitable for use in open-span applications (i.e. they may only be used at or near points of primary support). In all cases, the design of the beam and its connections, including the specification and number of bolts, must be determined by a qualified structural engineer.

### Flange and web plate splice

This is the most commonly used engineer-specified splicing method (Fig. 6.9). The splice is built using flange and web plates with high-tensile bolt fastenings. The disadvantage with this sort of splice is that plates and bolts project above and below the beam flanges. Any such projections should be taken into account relative to ceiling heights and floor clearances.

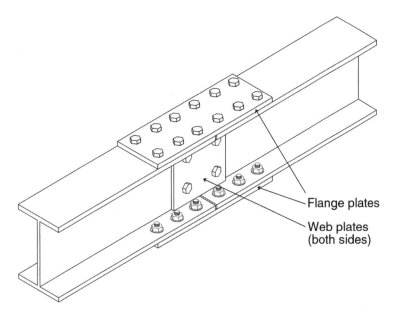

Flange plates

Web plates
(both sides)

**Fig. 6.9**  Flange and web plate splice.

## End plate beam splices

Connection is achieved via end plates that are shop-welded to beam ends (Fig. 6.10a). As with the flange/web splice described above, plate and bolt projections above and below the flange surfaces must be considered. A variant approach, sometimes called a butt splice, provides a seamless beam-to-beam connection with no upward or downward bolt/plate projections (Fig. 6.10b). In both examples, the integrity of the splice is dependent on the use of high-strength friction grip (HSFG) bolts (see *Bolted connections*) and these must be correctly tensioned (Fig. 6.10c).

## Splice box

An alternative to conventional bolt and plate splice connections, the splice box takes the form of a prefabricated heavy-duty sleeve that slides over the ends of the beams: beam ends are then brought into contact within the box and the connection completed with pinch bolts (Fig. 6.11a). Splice boxes can be purchased off the shelf and allow rapid installation of steelwork.

## Inline box

Using inline boxes, it is possible to insert between walls a length of beam that is marginally shorter than the gap it must span. A short cantilever stub is fixed to the supporting wall, with the two-part inline box forming a sleeved and bolted connection between the wall stub and the end of the main beam (Fig. 6.11b). This is a shear connection and is typically used within 150 mm of a load-bearing wall or post; it is not intended for open-span use. Like the splice box described above, the inline box can be purchased off the shelf and allows rapid installation.

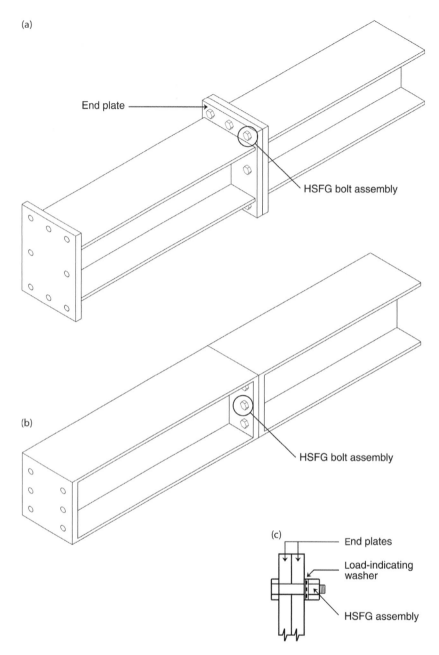

**Fig. 6.10**   (a) End plate beam splice, (b) Butt splice, (c) HSFG assembly.

## PFC bearing

With this arrangement, the main beam itself does not penetrate the supporting walls: final support is provided by a pair of PFCs at each end of the beam (Figs 6.12a and b). These are accommodated within the web of the main beam and bolted through. This approach eliminates some of the problems associated with positioning beams between existing masonry walls.

(a)

(b)

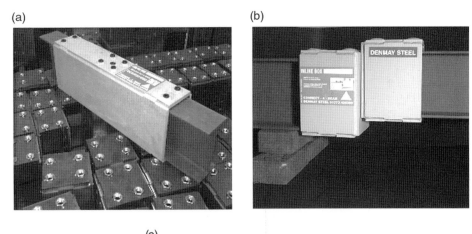

(c)

**Fig. 6.11** CONNECT-A-BEAM steel beam joining devices. Courtesy Denmay Steel. (a) Splice box, (b) Inline box, (c) Tee box.

## Beam-to-beam connections

Steel beams may be joined at an angle by means of angle cleat connections. The design and specification of such connections is, of course, the role of the structural engineer. Web cleats may be shop-welded to both sides of the web of the secondary beam. One or both of the secondary beam flanges are notched to create a sound connection between the beams (Fig. 6.13).

## BOLTED CONNECTIONS

Bolted connections are used in a range of structural applications. At one end of the scale, mild steel bolts may be used in the fabrication of simple timber-to-timber components such as trimmers and header beams. In situations such as these, the strength of bolts is unlikely to be a critical factor provided they are of adequate dimensions and appropriately spaced.

The opposite end of the spectrum is represented by bolted connections in beam splices. For example, the design of bolted connections in an end plate beam splice (described above) would include the specification of HSFG bolt assemblies. It should be noted that proof loads for HSFG assemblies are between two and three times greater than those for

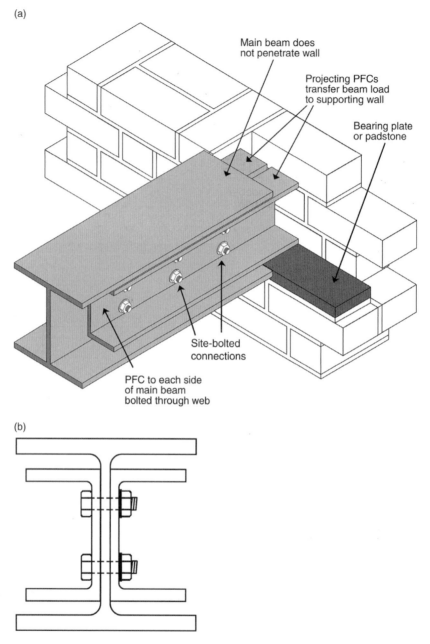

(a)

Main beam does
not penetrate wall

Projecting PFCs
transfer beam load
to supporting wall

Bearing plate
or padstone

Site-bolted
connections

PFC to each side
of main beam
bolted through web

(b)

**Fig. 6.12**   (a) PFC bearing, (b) PFC bearing – section.

black (mild steel) bolts. However, HSFG and black bolt assemblies are similar in appearance, and the ability to correctly identify them is therefore of considerable importance. The following notes are provided to clarify the use of various common bolt configurations.

*It is important to note that the finish or colour of a bolt does not provide a reliable indication of its strength or function. 'Black' bolts, generally the weakest used in structural*

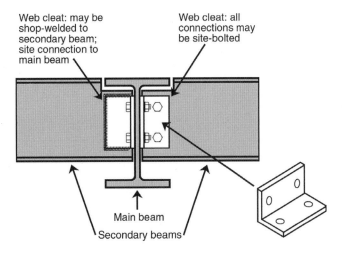

Web cleat: may be shop-welded to secondary beam; site connection to main beam

Web cleat: all connections may be site-bolted

Main beam

Secondary beams

**Fig. 6.13** Cleat connections.

**Table 6.2** Grade 4.6 bolt assemblies.

| | |
|---|---|
| Identification | Bolt head stamped 4.6 |
| Material | Mild steel |
| Applications | Timber-to-timber and timber-to-steel |
| Clearance hole | Generally bolt diameter + 2 mm |
| Proof load (M16) | 34.8 kN |

*connections, are not always black. Conversely, HSFG bolts (the strongest) may have a black phosphate finish.*

## Grade 4.6 bolts

Grade 4.6 bolts and nuts manufactured from mild steel are often referred to as 'normal strength' or 'black' bolts. Grade 4.6 bolt assemblies may be used in timber-to-timber and timber-to-steel connections (Table 6.2).

It is necessary to use washers both behind the bolt head and behind the nut when making timber connections. In addition, it is often necessary to use toothed plate connectors between timber elements. The diameter of washers in timber connections should be three times bolt diameter and the thickness of washers 0.25 of bolt diameter. Clearance holes are generally diameter + 2 mm. Thus, an M10 bolt would require a 12 mm clearance hole and 30 mm washers. Note that, when tightened, at least one complete thread should protrude beyond the nut.

## Grade 8.8 'high-strength' bolts

These are sometimes described as high strength or hexagon structural bolts (Fig. 6.14a). They are manufactured from high-tensile steel and are used in a wide range of structural applications including steel-to-steel connections (Table 6.3). Note that high-strength bolts are not the same as HSFG bolts.

(a)                              (b)                              (c)

(d)

**Fig. 6.14**   (a) Grade 8.8 bolt, (b) HSFG bolt, (c) HSFG load-indicating washer, (d) HSFG flat washer.

**Table 6.3**   Grade 8.8 bolt assemblies.

| | |
|---|---|
| Identification | Bolt head stamped 8.8 |
| Material | High-tensile steel |
| Applications | Timber-to-timber, timber-to-steel, steel-to-steel |
| Clearance hole | Design dependent |
| Proof load (M16) | 89.6 kN |

**Table 6.4**   HSFG bolt assemblies.

| | |
|---|---|
| Identification | *Bolt head*: three radial lines at 120° |
| | *Nut*: three circumferential arcs |
| Material | High-tensile steel |
| Applications | Steel-to-steel connections |
| Clearance | Design dependent |
| Proof load (M16) | 92.1 kN |

## HSFG bolt assemblies

These bolt assemblies are used in structural steel joints where high clamp forces are required to generate friction between mating surfaces (Fig. 6.14b). A typical application would be in the formation of bolted joints in beam splices where beam sections are connected via end plates. Like other high-strength bolts, HSFG bolt assemblies are produced from high-tensile steel (Table 6.4). However, HSFG and conventional 'high-strength' bolts should not be confused: HSFG bolt heads and nuts are wider across flats than grade 8.8 bolts of the same diameter.

Shank tension is critical in applications where HSFG assemblies are used, and load-indicating washers (direct tension indicators) are commonly used to ensure that minimum specified shank tension is achieved in the process of assembly (Fig. 6.14c). These washers have protrusions that are flattened progressively as shank tension increases to predetermined levels. Non-indicating flat washers for use in HSFG assemblies are identified by nibs (Fig. 6.14d).

## Toothed plate connectors

Toothed plate connectors, sometimes called timber connector plates, are used in structural timber-to-timber connections where two parallel lengths of timber are required to act in unison. Toothed plate connectors increase the effectiveness of bolted timber joints by improving stress distribution within the joint and by reducing the tendency of timber to shear parallel to the grain when loaded. Common applications include the formation of trimmers, trimming joists and double joists for partition or bath support (see Fig. 7.13).

## TIMBER TO MASONRY CONNECTIONS

It is frequently necessary to fix new timber elements such as sole plates or wall plates to existing masonry walls as part of a loft conversion. As with any structural connection, the choice of fixing system is influenced by the loads to which the new elements will be subjected.

The condition of the masonry substrate must also be taken into consideration when selecting a fixing system. The poor mechanical properties of some traditional stock bricks and the relative weakness of lime-based mortars will influence the choice of anchoring system. The effectiveness of the chosen anchoring system is also influenced by factors including embedment depth, edge distance, the nature of the bond between masonry and fixing, and to a lesser extent the shear/tension characteristics of the fixing itself. Timber to masonry anchoring methods include:

- Tension straps
- Expansion bolts
- Chemical anchoring

## Tension straps

Generally in 5 mm (heavy duty) or 2.5 mm (light duty) galvanised steel (Fig. 6.15a). Standard width 30 mm, lengths usually from 100 to 3600 mm in 100 mm increments. Tension or restraint straps are used for fixing wall plates and may also be used for fixing sole plates to wall heads. The disadvantage with using straps at wall plate level in loft conversions is that the leg of the strap (unless of minimal length) would generally extend through the existing ceiling into the room below and would consequently disturb wall finishes.

## Expansion bolts

Expansion bolts provide an effective method of fixing timber to masonry, but only if the masonry is in good condition (Fig. 6.15b). Friable stock bricks with weak traditional

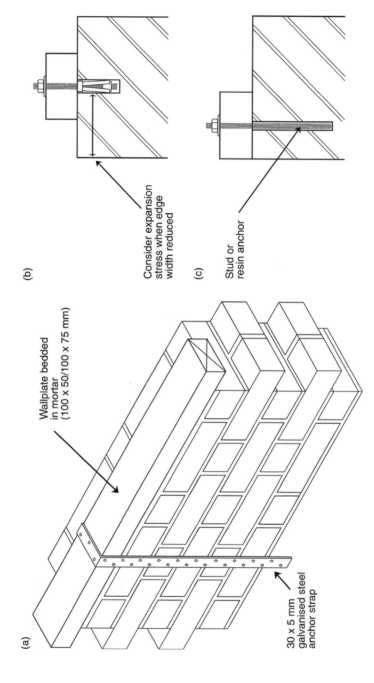

(b)

Consider expansion
stress when edge
width reduced

(c)

Stud or
resin anchor

Wallplate bedded
in mortar
(100 x 50/100 x 75 mm)

30 x 5 mm
galvanised steel
anchor strap

(a)

**Fig. 6.15** Timber to masonry connections. (a) Strapping, (b) Expansion, (c) Chemical fixing.

mortar are not a satisfactory substrate. Equally, expansion bolts are best used away from the edges of masonry where expansion stress can cause brickwork to fail locally.

## Chemical anchoring

Chemical anchoring systems provide, in effect, a glued stud connection, and this approach overcomes some of the problems associated with mechanical expansion fixings (Fig. 6.15c). A clearance hole is drilled in the masonry, dust removed from the hole and reagents introduced. Proprietary anchors are available, but threaded bar, properly degreased, may also be used to create a stud. This system is well suited to applications where fixings are to be provided close to the edge of masonry walls.

Two types of chemical anchoring are used:

- Resin anchors (generally based on polyester resin)
- Epoxy resin anchoring systems (which do not expand and may be used in conjunction with stainless steel studs)

## DISPROPORTIONATE COLLAPSE

Requirement A3 of the Building Regulations is that buildings '... shall be constructed so that in the event of an accident the building will not suffer collapse to an extent disproportionate to the cause'.

Guidance concerning disproportionate collapse is set out in Approved Document A *Structure* (2004) and compliance is primarily achieved by providing effective connections between walls and floors. Specific provisions are dependent on Building Class and this, as far as dwellings are concerned, is defined in relation to the number of storeys in the building and the type of occupancy. Houses not exceeding four storeys are included in Class 1; single-occupancy houses of five storeys are included in Class 2A.

The most common sort of loft conversion (one that creates a new storey in an existing two-storey house) would generally not trigger the need for any remedial action elsewhere in the building under the guidance on disproportionate collapse because the building would remain in the Class 1 designation. The same would also apply, usually, to an existing three-storey single-occupancy dwelling that becomes a four-storey dwelling as a consequence of a loft conversion. The conversion, of course, would need to conform to Class 1 robustness.

However, when building work creates a five-storey house (i.e. the conversion of the roof space in an existing four-storey house), the Building Class of the entire structure changes from 1 to 2A. In some cases, this might lead to the building being less satisfactory in relation to Requirement A3 than it was before alteration, and the need for remedial action may be triggered. Attention is drawn to the following points:

### Basement storeys

Unequivocal guidance on whether basement storeys should be included in the storey total is not provided in Approved Document A. Fig. 6.16 indicates a configuration in which a

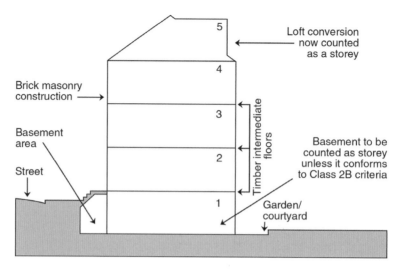

**Fig. 6.16**  Five dwelling: common configuration.

basement would generally be considered to constitute a storey for the purposes of the guidance. Note that while a basement storey *might* be excluded from consideration if it conforms to certain robustness criteria, this is unlikely to be the case in the mode of construction indicated here.

### Status of loft conversions

Requirement A3 (2004) applies to all buildings without limitation and a loft conversion is now considered to constitute a storey in its own right.

### Change of Building Class – remedial measures for 2A dwellings

Conversions that lead to the creation of a five-storey building are most likely in urban terraced dwellings constructed in the eighteenth and nineteenth centuries. Houses of this sort, constructed from either brick or stone masonry, are generally found only in the larger British towns and cities. As a rule, systematic anchorage between walls and floors was not provided in buildings of the period. Ties between walls and floor joists running parallel to them are likely to be absent; the effectiveness of existing 'natural' ties formed by wall-supported timber joists will vary depending on the nature of the junction and the condition of both masonry and timber.

The implication of the guidance is that Class 2A standards should be applied to the *entire* building when a loft conversion leads to the creation of a five-storey dwelling. This would require the retroactive installation of an anchoring system or horizontal ties capable of conforming to new-build standards, with straps at 1250 mm spacing between suspended floors and walls (including internal walls) throughout the building. Note that an alternative approach, based on a collapse-resistant support system for the conversion only, is sometimes acceptable.

# 7  Floor structure

Ceiling joists are seldom capable of supporting occupational floor loads. So when a loft is converted, it is usually necessary to construct an entirely new floor with joists supported by beams or by existing load-bearing walls (see also Chapter 6).

The floor has two primary structural functions. It must first be able to support its own dead load and other designed loads. In addition, it must also be able to support the loads imposed on it by users of the building and their possessions without undue deflection.

Ordinarily, determining loads for timber intermediate floors in dwellings, other than loft floors, is straightforward. Unless concentrated loads (for example, those caused by trimming members) or line loads (caused by partitions of more than 0.8 kN per linear metre) are factors, there is generally no requirement to provide structural engineering calculations for the joists themselves.

Selection of joists appropriate to the loading conditions may be made by reference to published span tables. For single-occupancy dwellings of up to three storeys, the current Approved Document A *Structure* (2004) refers readers to *Span tables for solid timber members in floors, ceilings and roofs for dwellings* (published by TRADA Technology Ltd). Note that this document has been superseded by *Eurocode 5 span tables: for solid timber members in floors, ceilings and roofs for dwellings* (TRADA Technology Ltd).

## ROLE OF THE CONVERSION FLOOR

Loft conversion floor designs are generally more complex than those of conventional intermediate floors. There are two reasons for this. One is that floor joists are generally supported by beams, rather than by walls directly. The specifications of structural elements such as beams must always be justified by calculation.

The other element of complexity concerns the loading to which the new floor may be subjected. Depending on the arrangement of the conversion, the design of the new floor may need to take into account a number of structural loads additional to normal dead and imposed loads. These might include:

- *Roof loads*: for example, where existing purlins are removed and a new purlin wall transmits a proportion of the roof load to the new floor structure.
- *External wall loads*: for example, the loads from new external stud walls including any portion of the roof supported by them.

*Loft Conversions*, Second Edition. John Coutts.
© 2013 John Coutts. Published 2013 by Blackwell Publishing Ltd.

■ *Lower floor ceiling loads*: for example, where ceiling binders are removed, the new floor must provide intermediate support for the ceiling structure in the rooms below.

In cases where roof and wall loads will be supported by the floor joists, rather than directly by elements of structure such as beams or existing masonry walls, the selection of joists must be justified by calculation.

The new floor may also need to be capable of performing two additional structural functions:

■ *Tying*: the new floor may be required to tie-in one or more roof slopes either directly, by connecting existing rafters to the new joists, or via intermediate elements linking the rafters to the new floor structure. This applies particularly where ceiling joists are removed below the new floor to provide increased headroom. Effective tying counteracts outward thrust generated by the roof slope, or slopes. An inadequately restrained roof slope will tend to spread the walls that support it, but as with many other forms of structural failure, it is not always immediately apparent and may only manifest itself over a number of years.

■ *Lateral restraint*: the new floor *may* be required to provide lateral restraint for the walls of the building. In the case of a conversion that creates a new storey in an existing two-storey house, it is generally assumed that the original roof structure, with appropriately fixed wall plates, rafters and ceiling joists, provides adequate lateral restraint, at least in the direction of the existing ceiling joists. However, as with tying (above), any alteration of the ceiling joist/rafter foot/wall plate configuration may reduce the ability of a wall to resist movement at right angles to its plane. In addition, the floor will be required to provide restraint where a new gable wall is raised.

In addition to its load-bearing functions, it is of equal importance that the floor (including the ceiling beneath it) be capable of providing the following:

■ *Fire resistance*: the new floor in the loft conversion of a single family dwelling, and other intermediate floors, must provide appropriate fire resistance. In single family dwellings, a 30-minute standard is required.

■ *Sound resistance*: the floor of the conversion must provide resistance to the passage of sound.

■ *Moisture resistance*: in the case of bathrooms, kitchens and utility rooms, any board used as flooring should be moisture resistant.

## ELEMENTS OF LOFT CONVERSION FLOOR DESIGN

The availability of headroom is a key factor in most loft conversions. Because planning considerations mean that raising the ridge height of a roof is usually not feasible, the overall thickness of the floor structure must be carefully considered in order to create reasonable room heights. A number of strategies are adopted to minimise the impact of floor thickness on available headroom. These include:

■ *Minimising joist span.* Deeper joist sections are required on extended spans, and it is generally prudent to limit clear spans as far as possible. The use of intermediate structural support, which could be provided by beams or perhaps by suitable internal walls beneath, may be considered. Note that the maximum span of any floor supported by a wall is 6 m (Approved Document A *Structure*, 2C23).

- *Position of new floor joists.* To optimise available headroom, new floor joists are generally run parallel to the existing ceiling joists, which are retained. This is usually achieved either by partially suspending the new joists from support beams (underslung joists) or by packing them up off existing wall plates. Note that the spacing of the new floor joists will be influenced by the position of the existing ceiling joists – see note on joist spacing below. In traditionally built houses (certainly pre-1950) with cut roofs, ceiling joist/ rafter spacing is not necessarily equal.

- *Dropping lower floor ceilings.* This approach may be used to gain headroom where clearances are particularly limited: the lower floor ceiling and supporting ceiling joists are removed and the new conversion floor structure inserted. This approach has the disadvantage of reducing ceiling height in the rooms immediately below the conversion and causes considerable disruption to the occupants of the building. It also requires a thorough assessment of the function of the existing pitched roof structure: as noted above, ceiling joists perform an essential tying function in most roofs and they should not be removed unless an alternative means of tying is provided. An alternative approach might include full restraint of rafters by a suitably designed ridge beam and purlins, and the introduction of new low-level ties.

## Room height in the conversion (headroom)

As noted in Chapter 5, there is no longer a stated minimum floor-to-ceiling height for rooms but it is generally accepted that a minimum of 2.3 m be maintained over the main portion of the conversion. Rooms with ceilings below this height are somewhat oppressive.

Proposals for reduced floor-to-ceiling heights within rooms should be carefully considered in relation to any downward projection from a ceiling. Where this is the case, it would be reasonable to consider guidance included in Approved Document K (AD K) *Protection from falling, collision and impact* and to ensure that the clearance between any such projection and floor level exceeds 2 m. Headroom on stairs and landings is considered at the end of the chapter.

## METHODS OF SUPPORT FOR FLOORS

Floor joists in a conversion may be supported in two ways:

*Beam support*: new floor joists are supported by new beams which transmit loads to existing walls or other suitable load-bearing structure. The ends of floor joists are either fixed into beam webbing or are supported by hangers nailed to timber bearers or packers that are bolted to the beam.

*Wall support*: new floor joists are supported directly by existing walls. Joists may be fixed to the tops of walls (on wall plates), or fixed to the face of walls (with joist hangers).

Depending on circumstances, a combination of beam and wall support may be used.

## Beam-supported floors

This is the most commonly used method of supporting the new floor in a loft conversion. To reduce the overall thickness of the floor structure, beams are set at right angles to existing ceiling joists, thus allowing new floor joists to be underslung between them. It is usually necessary to remove ceiling binders where this is the case (see *Binders*). A 25 mm clearance should be allowed between the underside of beams and the existing ceiling joists to eliminate the risk of damage caused by deflection.

Beams can be used to support floors in a number of different ways. The approach adopted will depend on the position of new floor joists relative to the beam. There are two principal approaches:

- New floor joists underslung from beams
- New floor joists into beam webbing

A third approach, based on additional beam support for existing ceiling joists, is also described.

### *Underslung floor joists*

Floor joists are partially suspended from supporting beams, allowing them to occupy the void between existing ceiling joists to maximise headroom in the conversion (Fig. 7.1).

The specification of appropriate joist hangers is of considerable importance in this application. Hangers designed for underslung support are produced by most of the major builders' metalwork manufacturers. These are often generically described as 'long-leg' or 'extended-leg' hangers, and a number of proprietary versions are available. All work

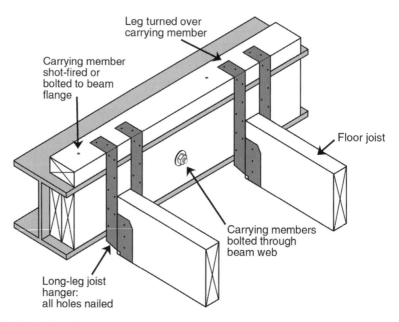

**Fig. 7.1**  Underslung floor joists.

similarly to conventional timber-to-timber hangers, the chief difference being the length of the legs, which are designed to wrap over the carrying member. Note that the safety and stability of floors supported in this way are almost entirely dependent on correct nailing and wrap-over, and it is therefore essential that the manufacturers' nailing schedules be adhered to – in most cases, all nail holes should be used.

Long-leg hangers cannot generally be fixed directly to steel beams and it is therefore necessary to fix timber carrying members or packers into the web and a bearing plate to the top flange of the beam to permit nailing. Methods of fixing bearers and packing timber to steel beams include bolting, shot-firing and the use of self-drilling screws.

The amount of joist downstand that may be achieved relative to the supporting beam depends on the design of the hanger: manufacturers' technical guidance must be followed. Typical maximum downstand for proprietary hangers is generally between 75 and 100 mm.

However, a careful assessment of relative downstand should be made where existing ceiling joists are to be retained in proximity to new floor joists: in some cases the leg flange at the base of the hanger will foul the adjacent ceiling joist.

### Floor joists into beam webbing

This configuration is generally used where the beam must be incorporated within the structural thickness of the floor, for example, where the floor, and perhaps the lower ceiling surface as well, must continue uninterrupted above and below the beam (Figs 7.2a and b). The cut end of the joist is accommodated within the web of the beam (often a universal column in this application) and supported on its bottom flange. Timber blocking should be fixed between joist ends in the web to limit lateral and torsional movement.

It is generally necessary for the joist to be scribed into the webbing with this arrangement. For example, where the joist is required to support both floor and ceiling materials, both the top and bottom surfaces of the joist should be proud of the flanges (as indicated in Fig. 7.2) to accommodate timber shrinkage that might otherwise draw finishes into direct contact with the beam. It would normally be necessary to provide calculations to justify significant notching of joists at their bearing points.

This approach may be used to support new floor joists above existing ceiling joists within the roof space, although obviously with a loss of headroom. It is also used where the ceiling of the lower floor is 'dropped' to gain headroom.

### Additional support for existing ceiling joists

Although this approach depends on support provided by beams, it is quite different in principle from the approaches described above: beams are fixed at appropriate intervals *beneath* and at right angles to the ceiling joists to reduce their effective span as illustrated in Fig. 7.3. Thus supported, the existing ceiling joists are then used as floor joists in the conversion. The advantage with this approach is that it allows headroom to be maintained in the conversion and considerably reduces the amount of timber required to construct the floor because new floor joists are not required.

This system is seldom adopted for a number of reasons, not least the disruption it causes to the occupants of the building. The following practical points would require consideration:

(a)

(b)

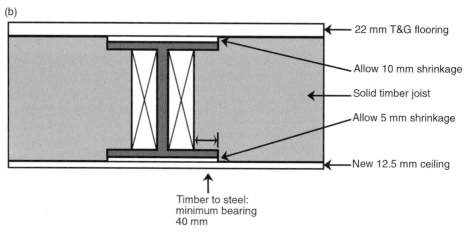

**Fig. 7.2**   Floor joists into beam webbing. (a) Floor beam and joists in situ. (b) Floor beam and joists – section.

- *Existing ceiling joists and supports*: the adequacy of wall plates and any internal load-bearing walls must be assessed and the sectional dimensions, span and spacing of ceiling joists noted. In most cases, these joists will not be grade marked, and it will be necessary to make a judgement on their strength class based on visual assessment.
- *Buildability*: the practical constraints of handling beams should be considered particularly in relation to obstacles such as partition walls through which they must pass. In order to make steel beams manoeuvrable, it would generally be necessary to

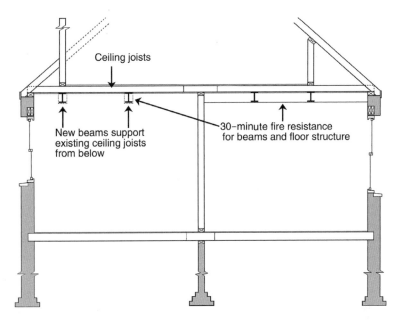

**Fig. 7.3**   Additional support for existing ceiling joists.

split them into sections with spliced connections designed by a structural engineer (see Chapter 6).

- *Relative beam position*: beams are installed at ceiling level in the floor immediately below the conversion. Ceiling materials, such as plasterboard or lath and plaster, must be cut away to allow direct contact between the beam and ceiling joists. The position of beams is influenced by the requirement to limit the span of the joists above and by the presence of obstructions: it is not permissible to fix structural elements, such as beams and joists, directly to chimney breasts or flues. In many cases, the relatively close spacing of the supporting beams will limit space for staircase access to the conversion and it will be necessary to introduce additional beams to trim a suitable floor opening.
- *Effect on lower floor rooms*: beams installed at ceiling height on the lower floor will clearly reduce floor-to-ceiling height for rooms and circulation spaces at the lower level.
- *Fire resistance of beams*: because floor beams must be protected from fire, it will be necessary either to clad them with an appropriate fire-resisting material or to construct an entirely new ceiling beneath them offering 30-minute fire resistance (see *Floor fire resistance* below). The use of intumescent paint may also be considered.

## Wall-supported floors

Most traditionally constructed houses and a high proportion of contemporary dwellings include structural elements that have the potential to provide support for a new floor (Fig 7.4). These include existing wall plates set on opposing external load-bearing walls

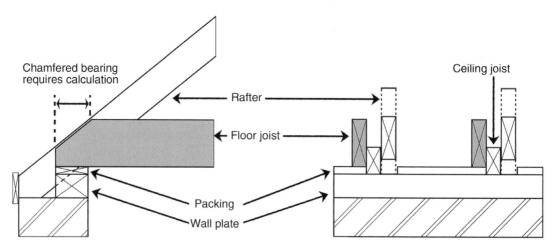

**Fig. 7.4**   Wall-supported floor joists.

and, particularly in pre-war dwellings, a central spine wall that in many cases is configured for a proportion of roof loading.

A thorough structural assessment must be carried out before undertaking the design of such a conversion. Particular attention should be paid to the following:

- *External walls*: wall plates must be adequately fixed to load-bearing walls, be properly bedded in mortar and be in good condition. The position of windows, doors and other openings lower down the wall must be considered.
- *Internal walls*: the spine wall, or other internal walls proposed for use, must be of load-bearing construction. In addition, it will be necessary to check that such walls are adequately restrained, are in true vertical continuity between floors and have adequate footings.
- *Joist/wall plate relationship*: if it is necessary to notch or chamfer the top of the joists to accommodate them beneath the pitch of the roof, this will generally need to be justified by calculation (Fig. 7.4).
- *Lintels*: where it is proposed to support joists on a wall plate or in masonry above a lower floor window, it would generally be necessary to prove the structural adequacy of the lintel, perhaps by exposing it, and to replace it if necessary. An alternative approach would be to trim around the window opening (Fig. 7.5).
- *Bays*: bay structures may have limited load-bearing capacity and should be treated with caution. As with lintels (above), trimming could be considered as an alternative means of support.
- *Buildability*: in addition to the structural factors outlined above, the potential difficulties associated with providing adequate fixings for joists in often confined eaves spaces should be considered (Fig 7.4).
- *Span*: guidance provided in Approved Document A *Structure* limits the span for any floor supported by a wall to 6 m, with span measured from the centre of the bearing.
- *Structural clearance*: floor joists should be set on packing to provide 25 mm of clearance from the existing ceiling structure.

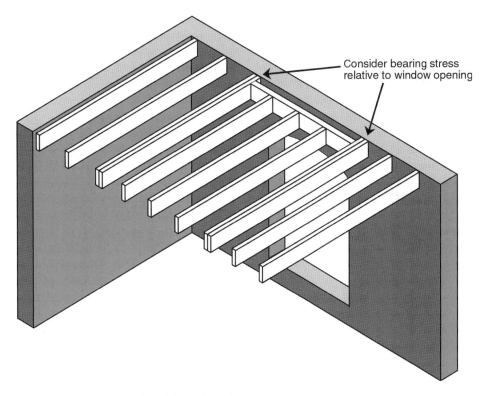

Consider bearing stress
relative to window opening

**Fig. 7.5**   Wall-supported floor joists: trimming.

## FLOOR JOIST SELECTION

Floor joists used in domestic loft conversions are generally produced from solid timber that is graded according to strength. This is often generically described as softwood carcassing timber. Timber of strength classes C16 or C24 (stronger than C16) is generally used. It is also possible to use timber I-joists in this application.

Unlike beams, there is no requirement to provide structural engineering calculations to justify the size of solid timber floor joists (unless additional loads must be considered – see beginning of chapter). As noted earlier, sizing may be determined by reference to *Eurocode 5 span tables: for solid timber members in floors, ceilings and roofs for dwellings* (TRADA Technology Ltd).

In all cases, however, it is important that timber of the appropriate strength class is specified – and used – during construction. Graded timber is marked to indicate conformity with its particular strength class and it is important that any timber used in structural applications displays such a mark in order to demonstrate compliance. BM TRADA Certification grade marks are illustrated in Figs 7.6a and b.

### Joist spacing

In buildings constructed until about 1970, spacing generally conforms to imperial dimensions, with typical centre-to-centre measurements of 14″ (approximately 356 mm),

(a)                                    (b)

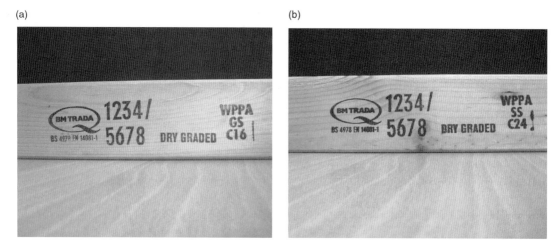

**Fig. 7.6**  BM TRADA Certification timber grade marks. Courtesy TRADA Technology Ltd. (a) C16, (b) C24.

**Table 7.1**  Commonly used joist sections (sawn).

| | |
|---|---|
| 47 × 125 mm | 75 × 125 mm |
| 47 × 150 mm | 75 × 150 mm |
| 47 × 175 mm | 75 × 175 mm |
| 47 × 200 mm | 75 × 200 mm |
| 47 × 225 mm | 75 × 225 mm |
| 47 × 250 mm | 75 × 250 mm |

16" (approximately 406 mm) or 18" (approximately 457 mm). As noted earlier, however, the spacing of ceiling joists and rafters in traditional cut roofs is not necessarily regular; indeed, it was often a somewhat haphazard affair with spacing varying considerably.

## Timber supplies

Whatever the technical basis for joist selection, it should be noted that the stock availability of given section sizes is also a determining factor. For example, timber with a breadth of 47 mm is the most widely used and represents a high proportion of timber imported to the UK (Table 7.1). By contrast, 50 mm timber is rather less common.

Length is also a determining factor. Generally, timber is sold in 300 mm increments (sometimes described as metric 'feet'), with a typical range from 1.8 to 7.2 m. Any specification that requires a special order is likely to add to costs and may take longer to obtain.

## Machined (regularised) joist sections

The section sizes quoted by wholesalers represent the nominal dimensions of timber as it rolls off the producer's sawmill. However, timber should be machined (regularised) on width (depth) on both opposing faces to provide an even surface for floors and ceilings. Published span tables take this additional processing into account. Note the conventions

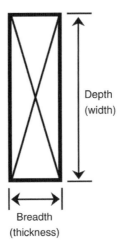

Depth
(width)

Breadth
(thickness)

**Fig. 7.7**   Joist dimensions.

for expressing the major and minor dimensions of timber sections, and particularly the use of the word 'width' in this context (Fig. 7.7).

## Holes and notches in joists

It is usually necessary to cut into or drill through joists in order to accommodate cables and pipes that run at right angles to the direction of the joists. Note that the rules for the safe positioning of holes and notches are quite different from each other:

- *Holes*: these should be drilled at the centreline (neutral axis) of the joist (Fig. 7.8a). The diameter of holes should not be greater than 0.25 times the depth of the joist and they should not be less than three diameters apart, measured from centre to centre. The safe zone for holes starts 0.25 of span length away from the support and ends 0.4 times the span length away.
- *Notches*: the safe zone for notches lies between 0.07 and 0.25 times the span length away from the support (Fig. 7.8b). Notches should not be deeper than 0.125 times the depth of the joist.

Note that notches and holes should generally not be cut in rafters, purlins or binders unless justified by calculation.

## BINDERS

In traditional 'cut' roof construction, ceiling joists were generally fixed by nailing to wall plates at each end of their span and also to the feet of the rafters for which they provide a tie in the majority of cases. However, in order to reduce the sectional dimensions of the timber needed to support the ceiling, intermediate support is often provided by binders (sometimes called ceiling bearers, runners or hanging beams) that are fixed above, and

(a)

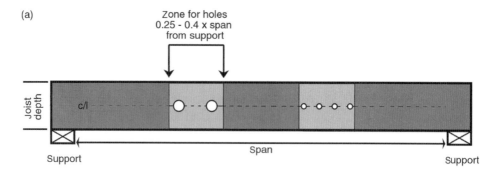

Zone for holes
0.25 - 0.4 x span
from support

Joist depth

c/l

Span

Support

Support

(b)

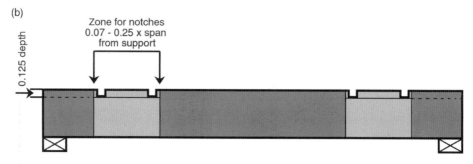

Zone for notches
0.07 - 0.25 x span
from support

0.125 depth

**Fig. 7.8**  (a) Hole positions (floor and ceiling joists), (b) Notch positions (floor and ceiling joists).

run at right angles to, the ceiling joists (Fig. 7.9). Binders are also found in TDA trussed roofs. Note that the binders used in modern trussed rafter roofs are generally configured to resist lateral loading.

Binders in cut roofs are supported at each end and sometimes at intermediate points on span, and act as beams as far as the ceiling structure is concerned. Two additional versions of the conventional binder are sometimes encountered. One takes the form of a section of timber fixed to the top of the ceiling joists broad side down. In this configuration, the 'binder' may not be supported at its ends. In this sense, it is not a true binder, but such an element does provide a degree of load sharing across the ceiling joists.

In another configuration, the binder may provide support not only for the ceiling joists but also for purlin struts. A binder of this sort (sometimes called a strutting beam) is subject to both roof and ceiling loads (see Fig. 9.4c). In order to gain headroom in the conversion, it is usually necessary to remove binders to allow new floor joists to be positioned between the existing ceiling joists. When binders are removed, a new way of providing intermediate support for the existing ceiling must be found. Where spacing allows, this may be achieved by positioning the new floor joists next to the ceiling joists: support for the ceiling joists may then be provided by strapping, screwing or nailing them directly to the new floor joists. Alternatively, existing ceiling joists may be strapped to the new floor support beams (Fig. 7.10d).

Strapping – using lengths of perforated galvanised steel band – is generally preferable to nailing or screwing. The strapped connection offers a degree of flexibility, and therefore any small deflection of the new floor joists is less likely to result in damage to ceiling

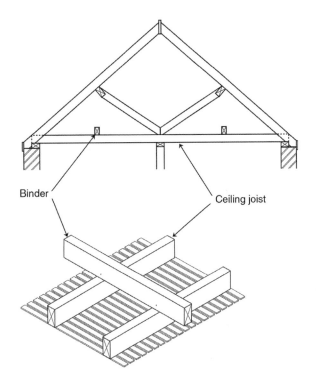

**Fig. 7.9**  Binders.

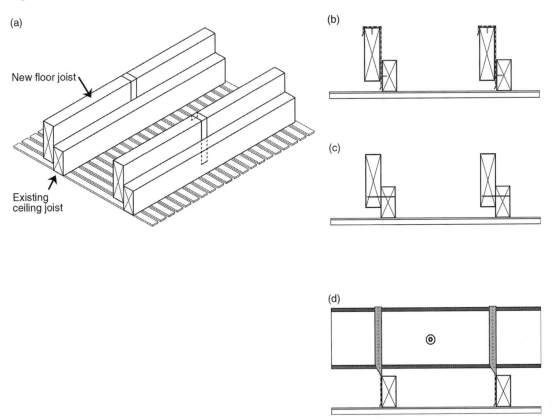

**Fig. 7.10**  Binder replacement. (a) Relative position of elements, (b) Strapping, (c) Nailing/screwing, (d) Strapping to new floor beam.

surfaces (Figs 7.10a and b). As noted above, the alternative is to nail or screw through the existing ceiling joists to the new floor joists (Fig. 7.10c), although this has the obvious disadvantage of creating a rigid connection.

Where straps or other connections are to replace binders, it would be reasonable to provide the new supports at the same point on span as the original binder on the grounds that this arrangement would be no worse than the original. It may be necessary to provide temporary support for the ceiling while binders are removed.

## New floor joist/existing ceiling clearance

It is practice to maintain a clearance of 25 mm between the underside of new floor joists and the existing lower floor ceiling. The primary reason for doing so is that it reduces the risk of damage to the ceiling caused by joist deflection.

In terms of buildability, there is also a strong argument for leaving a reasonable clearance between old and new floor elements. This applies particularly in older structures where it is unlikely that the ceiling will be level across the width of the building, and a degree of relative concavity is likely between supports.

## STRUTTING

Strutting, sometimes described as blocking or nogging, is fixed between floor joists (and flat roof joists) to counteract the timber's natural tendency to twist as it dries out. It is also effective in enhancing load sharing between joists. In addition, strutting helps to reduce vibration in floors.

Blocking should be provided at the supported ends of joist runs (e.g. between joist ends within beam webbing). Intermediate strutting is not generally required for spans less than 2.5 m. For joists with a span of more than 2.5 m, strutting is provided mid-span. Where the span exceeds 4.5 m, two rows of strutting should be provided with one at each third point along the span. Note that the breadth-to-depth ratio of joists may also influence the location of strutting.

Solid timber strutting, timber herringbone strutting and proprietary metal strutting may all be used in this application. However, in order to fix diagonal bracing such as herringbone and proprietary strutting systems, access is generally required from below as well as from above the joist. Unless the lower floor ceiling and its supporting joists have been removed as part of the conversion, and unless the joists are evenly spaced with a clear void, it is not possible to use these approaches and solid strutting should be used instead.

Solid struts, a minimum of 38 mm in breadth, should be less than the full depth of the joist to minimise the risk of disturbing floor or ceiling surfaces. Guidance is for a minimum of 0.75 depth. In a loft conversion where existing ceiling joists project into the void between the new floor joists, it is not always possible to fix strutting of minimum depth across the full width between the joists, and it is therefore necessary to notch the struts to accommodate the ceiling joist projection. Where this is necessary, it would be reasonable to provide a clearance of at least 25 mm between the strut notch and the ceiling joist to accommodate deflection (Fig. 7.11).

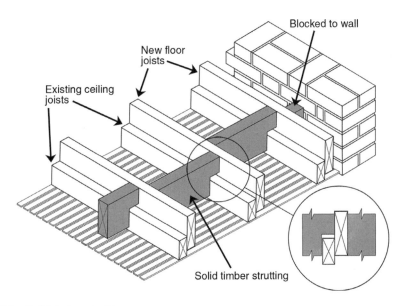

Blocked to wall

New floor joists

Existing ceiling joists

Solid timber strutting

**Fig. 7.11** Solid timber strutting.

## TRIMMING

Trimming members are introduced to provide support where joist runs are interrupted by projections or openings in the floor structure. In a loft conversion, there are typically two sets of circumstances where trimming is necessary:

- Around the stairwell opening to the new floor
- Around a chimney breast

A third set of circumstances where floor trimming may be used is where new floor joists must be supported by an existing masonry wall that contains an opening with a lintel of unknown strength. It would clearly not be acceptable to introduce a new loading directly above such an opening and it is therefore practice either to fix a lintel of known strength or to trim around the opening (Fig. 7.5).

Where primary support for the new floor joists is to be provided by steel beams, secondary beams may serve as trimming members for stairwell openings and projections such as chimney breasts. In the case of chimney breasts, the secondary beam may perform a dual role, providing both a trimming 'joist' for the floor and also support for a new ridge beam via a timber post (see also Fig. 6.1).

Conventional approaches to trimming using timber members are still widely used in floor framing (Fig. 7.12). There are typically four elements in a trimmed floor:

(1) *Bridging joists*: bridging or common joists run between their natural supports without interruption.
(2) *Trimmed joists*: bridging joists that are cut short to accommodate the opening or projection.

(a)

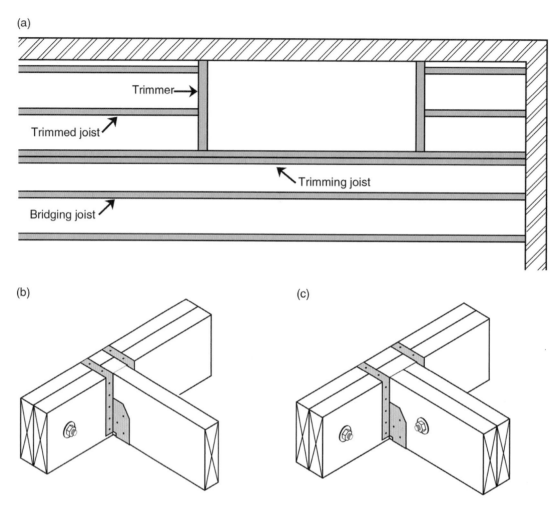

Trimmer⟶

Trimmed joist

Trimming joist

Bridging joist

(b)                                                              (c)

**Fig. 7.12**   Floor trimming. (a) Trimmed opening configuration, (b) Trimming joists to single trimmer, (c) Trimming joists to double trimmer.

(3) *Trimmer joist*: a joist that runs at right angles to, and provides support for, the cut ends of the trimmed joists.

(4) *Trimming joists*: trimming joists run parallel to the bridging joists and provide support for the trimmer.

In effect, the trimming and trimmer joists behave as beams. Because they are subjected to high point loads, it is necessary that they be made stronger than the trimmed joists they support. The requirement to provide a regular surface for ceilings and floors means that it is generally not practical to provide deeper joists and it is therefore practice to provide trimming members of increased breadth. Note that trimming joists must be fixed together with bolts and toothed plate connectors as illustrated in Fig. 7.13.

The traditional approach to sizing trimming and trimmer joists supporting no more than six trimmed joists was to make them 25 mm thicker than the bridging joists. However, under current practice it would be unwise to proceed on this basis and the dimensions of trimming members should be determined by calculation.

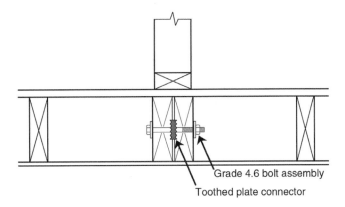

Grade 4.6 bolt assembly
Toothed plate connector

**Fig. 7.13** Double joists.

Connections between timber members within the trimmed structure are now usually made using metal joist hangers or fixing plates (Figs 7.12b and c). Traditional joints, such as tusk tenon and blind tenon joints, are relatively time consuming to make and their effectiveness is only achieved through craftsmanship of a high order.

In designing the conversion, it is clearly desirable to limit the extent of trimming as far as possible, particularly where large openings in the floor are required. For example, ensuring a parallel relationship between a new stairwell serving a loft and floor joists in the conversion will considerably reduce the need for trimming.

As noted above, it is often necessary to trim floor support structures around chimney breasts (Fig. 7.14). Guidance on clearances between combustible materials (such as timber joists) and flues is contained in Approved Document J. Note that there is a presumption against making any structural connections to chimney breasts.

## LATERAL SUPPORT BY FLOORS

Guidance in Approved Document A *Structure* is that floors above ground level should be fixed to walls (Table 7.2). This allows the floor to act as a horizontal brace. Lateral support of walls by floors is effective in restricting the movement of external walls. Internal load-bearing walls also require lateral support. Note that a new roof may also provide lateral support for walls.

Formalised guidance on lateral support of walls by floors and roofs is, in historical terms, a relatively recent development. Traditional modes of construction meant that walls supporting floors and ceilings often had a reasonable degree of intrinsic stability with floor and ceiling joists fixed to them providing lateral support. However, designed connections between floor joists and walls *parallel* to them (i.e. generally along flank walls) were not routinely provided.

Most loft conversions are, of course, carried out in traditionally constructed buildings. Where elements that already provide lateral support for existing walls are removed or altered, it is necessary that restraint be provided in some other way. This may occur, for example, if ceiling joists are removed (but note also their independent roof tying function described earlier).

The position is less clear as far as previously *unrestrained* existing external or compartment walls are concerned. Examples of unrestrained walls may be found frequently in the

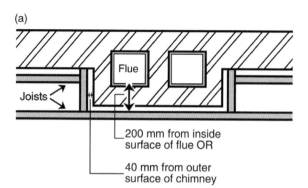

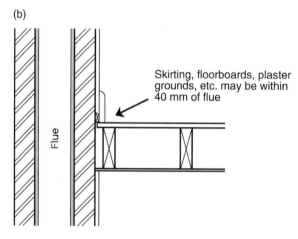

**Fig. 7.14**   Combustible materials: proximity to flue/chimney. (a) Trimming to chimney/flue – plan, (b) Proximity of minor elements – section.

**Table 7.2**   Lateral support for walls provided by floors and roofs.

| Wall type | Wall length | Lateral support required |
|---|---|---|
| Solid or cavity: external wall, compartment wall or separating wall | Any length | Roof lateral support by every roof forming a junction with the supported wall |
| | Greater than 3 m | Floor lateral support by every floor forming a junction with the supported wall |
| Internal load-bearing wall (not being a compartment or separating wall) | Any length | Floor or roof lateral support at the top of each storey |

case of flank walls of detached, semi-detached and end-of-terrace dwellings. It would clearly be good practice to provide connections between the new floor and such walls, but this is sometimes overlooked.

However, where the conversion involves the construction of a new wall (as in a hip-to-gable conversion) or the extension of an existing one (e.g. where a gable is extended to form a flank gable), the lateral restraint of those walls by the new floor should be considered. Fig. 7.15 illustrates methods of providing lateral restraint for cavity and solid masonry walls.

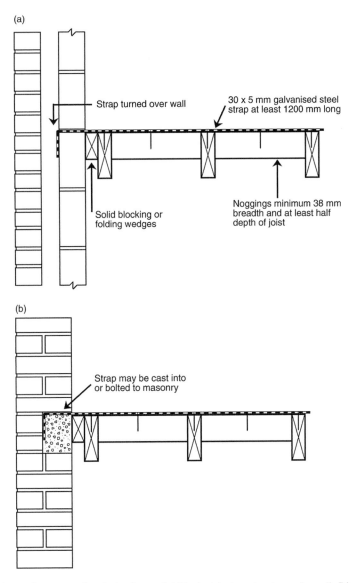

**Fig. 7.15**   Lateral support of walls by floors. (a) Restraint strapping in cavity wall, (b) Restraint strapping in solid masonry wall.

Where it is necessary for a new floor to provide lateral support, the guidance set out in Approved Document A *Structure* may be followed. However, Approved Document A lists a number of conditions where tension (restraint) straps need not be provided, including:

■ in the longitudinal direction of joists in houses of not more than two storeys, if the joists are at not more than 1.2 m centres and have at least 90 mm bearing on the supported walls or 75 mm bearing on a timber wall plate at each end; and
■ in the longitudinal direction of joists in houses of not more than two storeys, if the joists are carried on the supported wall by joist hangers in accordance with BS EN 845-1 of

the restraint type described in BS 5628: Part 1 and are incorporated at not more than 2 m centres; and

■ where floors are at or about the same level on each side of a supported wall, and contact between the floors and wall is either continuous or at intervals not exceeding 2 m. Where contact is intermittent, the points of contact should be in line or nearly in line on plan.

## FLOOR FIRE RESISTANCE

In order to protect occupants and structure from the effects of fire, guidance set out in Approved Document B *Fire safety* is that the new floor in a loft conversion should have a specified standard of fire resistance.

Note that the fire resistance of *all* intermediate floors in the dwelling, and elements of structure that support them, must be assessed and upgraded if necessary when a loft is converted. Normally, full 30-minute fire resistance is required for floors, but under certain circumstances (outlined in Chapter 4), a modified 30-minute standard is permissible at the first-floor level.

Without testing a given floor structure, it is impossible to ascertain precisely what level of fire resistance it would offer. However, a number of assumptions are made about existing floor structures. For example, it is assumed that a plaster on wood-lath ceiling of between 15 and 22 mm thickness *might* provide 20-minute fire resistance if it is in good condition (i.e. plaster nibs are intact and form an unbroken bond with the laths). A thicker ceiling would not necessarily perform any better and, indeed, might fail more rapidly. It is also assumed that plain-edge floorboards, or badly fitting tongue and groove boards, contribute little to overall fire resistance (Fig. 7.16).

Given that there is often a reluctance to modify existing ceilings for practical and sometimes aesthetic reasons, remedial measures are applied from above rather than below wherever possible. These measures might include the introduction of mineral wool pugging between floor joists. This contributes to fire resistance provided it is brought into close contact with the joists. In addition to this, overlaying plain-edge boards or badly fitting tongue and groove (T&G) flooring with either 3.2 mm hardboard or 4 mm ply will also enhance fire resistance.

In all cases, the ceiling must be imperforate in order for floor fire resistance requirements to be met. Recessed lighting is likely to compromise the integrity of the ceiling. In all cases where fire resistance is required, light fittings should be protected by appropriate fire hoods installed according to the manufacturer's instructions.

## Conversion floor (fire and sound resistance)

The new conversion floor must comply with guidance on fire resistance set out in Approved Document B *Fire safety*. It should also conform to guidance on sound resistance (Approved Document E *Resistance to the passage of sound*).

Fire and sound resistance are, to a certain extent, complementary functions. Ceiling finishes and mineral wool of appropriate specification possess both fire and sound-resisting properties; flooring (i.e. boarding) also contributes to fire and sound resistance.

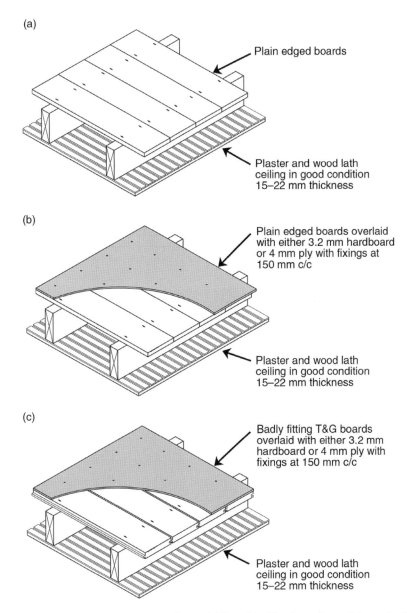

**Fig. 7.16**  Floor fire resistance: existing floors. (a) Possible 20-minute fire resistance (ceiling component only), (b) Possible 30-minute fire resistance, (c) Possible 30-minute fire resistance.

Fig. 7.17 indicates a possible configuration. Note that there is currently no requirement to retrospectively provide sound insulation for existing floors elsewhere in the dwelling.

## FLOOR MATERIALS AND FIXING

To allow access for first-fix plumbing and first-fix electrical work, and to facilitate inspection of the floor structure as part of the building control process, boarding is not generally immediately fixed in its final position after joists are installed. Instead, boards may be

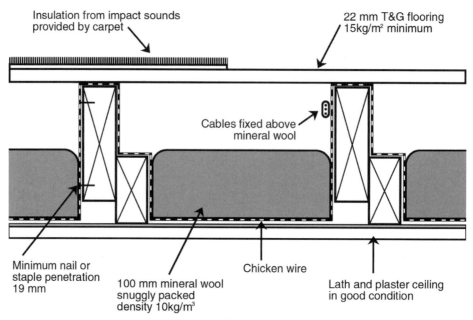

Insulation from impact sounds
provided by carpet

22 mm T&G flooring
15kg/m² minimum

Cables fixed above
mineral wool

Minimum nail or
staple penetration
19 mm

100 mm mineral wool
snuggly packed
density 10kg/m³

Chicken wire

Lath and plaster ceiling
in good condition

Note: recessed light fittings to be fitted with fire hoods

**Fig. 7.17**  Conversion floor: section.

fixed selectively until work has advanced to a point where it is practical to permanently fix the flooring. While fixing techniques vary depending on the type of flooring selected, there are three basic principles common to most boarding materials.

## Conditioning

Most modern composite materials intended for flooring have a high degree of dimensional stability provided they remain dry. However, many types of board may benefit from being conditioned, or acclimatised, by being laid loose and flat in situ before being fixed. In the case of softwood timber boards, a degree of shrinkage must be anticipated, even with heavy cramping, in all but the most highly seasoned timber.

## Staggered joints

If timber T&G boards are to be used, joints should be staggered: heading joints reduce load sharing within the floor and in the case of exposed timber boards, rows of heading joints are considered unsightly.

## Moisture and sound resistance

Note that in rooms where water may be spilled, such as bathrooms, utility rooms and kitchens, the guidance in Approved Document C *Site preparation and resistance to contaminants and moisture* (2004) is that flooring should be moisture resistant.

Boards should be laid with identifying marks uppermost to demonstrate compliance. If softwood boards are to be used, they should be no less than 20 mm thick and from a durable species as set out in BRE Digest 429. Alternatively, softwood boards should be treated with a suitable preservative. Note that, to conform to guidance set out in Approved Document E *Resistance to the passage of sound*, flooring should have a minimum mass of 15 kg/m$^2$.

## Fixing

In order to prevent creaking, flooring should be fixed with either ring shank nails or screws. The joints of tongue and groove panels should be glued. Note that conventional wire and cut nails have relatively poor resistance to withdrawal.

## T&G floor panels

Flooring-grade tongue and groove chipboard panels are now used in most applications, with a typical board size of 2400 or 2440 × 600 mm and a board thickness of 22 mm. The interlocking edges of these boards allow them to be laid with long edges running at 90° to the joists. However, where long edges meet walls, 50 × 75 mm support noggings should be provided to prevent the boards from sagging. Short edges bear on joists. The boards are staggered to increase rigidity, and joints should be glued with waterproof PVA (polyvinyl acetate) adhesive to reduce the risk of creaking.

## Timber floorboards

Tongue and groove timber boards may be used for flooring. Note that sanded and sealed T&G floorboards, while attractive, are best restricted to ground-floor use in the interests of impact noise reduction.

## STAIRS

Guidance on stair access is provided in AD K *Protection from falling, collision and impact*. The guidance is relatively straightforward and it includes a number of important concessions that are of assistance when loft conversions are carried out. Additional information on stair specification can be found in the BS 585 and BS 5395 series of standards.

## Headroom

Although there is no longer a minimum floor-to-ceiling height within rooms, AD K indicates that a minimum of 2 m headroom be maintained on stairs and landings (Fig. 7.18a). This should be considered with particular reference to any downward projecting elements (such as beams and purlins) which present an impact hazard.

In the case of loft conversions where it is not possible to achieve a 2 m clearance, AD K indicates that reduced headroom is permissible with 1.9 m to the stair centreline and a minimum of 1.8 m on one side only (Fig. 7.18b). Again, the position of downward projecting elements must be considered carefully.

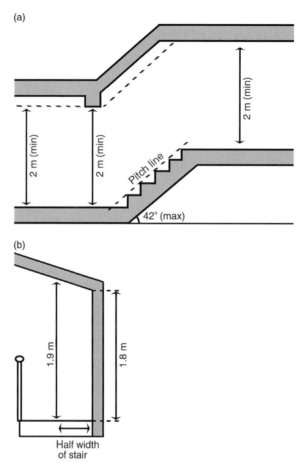

**Fig. 7.18**   Stair access: headroom. (a) General headroom requirements on stairs, (b) Reduced headroom (loft conversions only).

## Landings

Landings should be provided at the top and bottom of every flight and be at least as wide and long as the smallest width of the flight. A door opening directly onto a stair would not be acceptable – but the definition of 'stair' in this context is sometimes complicated by the rather complex hybrid arrangements of some conversions (see Appendix B, appeal reference 45/3/165).

ADK recognises the hazards presented by doors and doorways relative to stairs, and provisions for minimising risk are shown in Fig. 7.19.

## Stair configuration

The provision of stairs has important implications for the layout of both the conversion and the floor immediately below it. In some cases, the need to accommodate stairs will lead to a reduction of the amount of habitable space on the floor below and it is for the householder to decide whether the space gained by the conversion justifies the loss of a

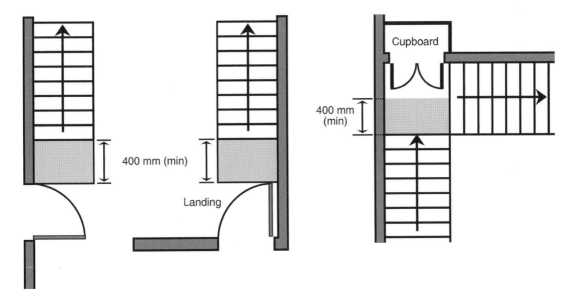

**Fig. 7.19**  Landings next to doors.

room, or part of one. However, a number of measures may be considered to limit the amount of space occupied by stairs. Bearing in mind the guidance on minimum pitch and headroom set out in AD K and fire safety guidance in Approved Document B, the following factors may be considered.

### Position of stair

To optimise the use of existing circulation areas such as landings, it is practice to run the new stair over the existing staircase where possible (Fig. 7.20). In order to provide appropriate headroom, the size of the ceiling opening required in most cases is only marginally smaller than the overall footprint of the stair.

### Stair form

A number of configurations are possible. For the purposes of loft conversion, it is assumed that the primary intention of the design process will be to limit the footprint of the stair and the size of the ceiling opening as far as possible. Fig. 7.21 illustrates five basic configurations. As a general principle, complex changes of direction should be made at the bottom of the flight where possible (Fig. 7.21b). While the arrangement illustrated in Fig. 7.21c would usually be acceptable, the provision of winders at the top of the flight introduces increased risk for users.

As already noted, landings must be provided at the top and bottom of every flight, and the width and length of every landing should be at least as great as the smallest width of the flight. The landing may include part of the floor of the building. For example, assuming that the stair illustrated in Fig. 7.21a is 700 mm wide, the clear space at the head and foot of the stair should be 700 × 700 mm (see also Fig. 7.19 with reference to potential obstructions).

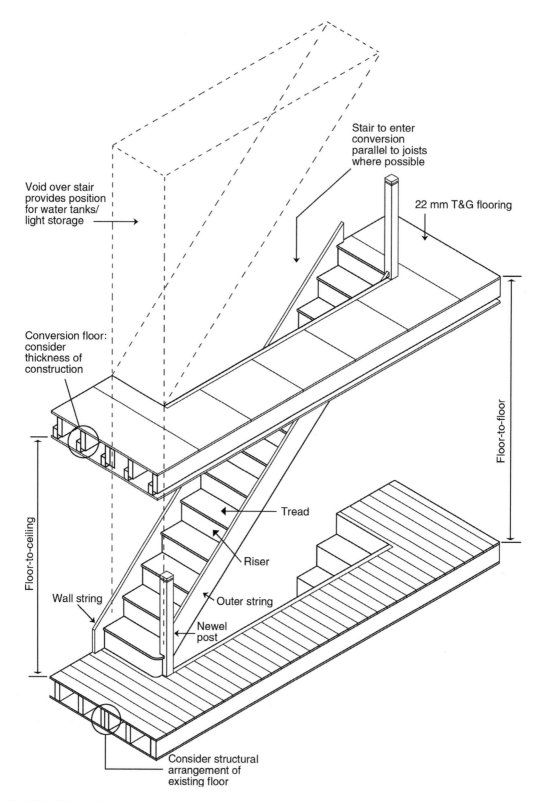

Fig. 7.20    Floor-to-floor configuration (balustrade omitted for clarity).

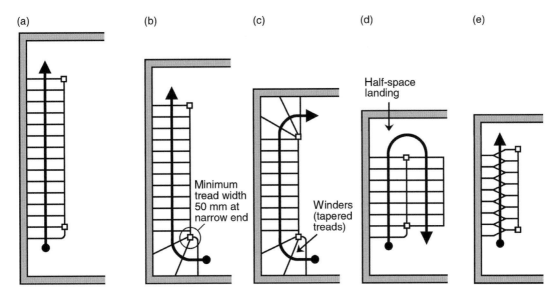

**Fig. 7.21**  Basic stair forms. (a) Straight flight, (b) Single winder, (c) Double winder, (d) Dog leg, (e) Alternating tread.

Guidance relevant to stairs contained in the Approved Documents (specifically, AD B *Fire safety* and AD K section 1, *Stairs and ladders*) is concerned only with safety. Under the guidance, it would be possible to construct a stair that would be entirely impractical for the purposes of moving furniture or large fittings between levels.

### Conventional stairs

AD K indicates a maximum pitch of 42° for conventional stairs. By virtue of the guidance, a rise of 220 mm (the maximum permitted) would allow a going not less than 245 mm. There is no regulatory minimum stair width. While 800 mm is preferred, 600 mm would constitute a reasonable minimum. In all cases, appropriate guarding must be provided, and the stair should have a handrail.

### Alternating tread stairs

These are sometimes described as space-saver stairs (Fig. 7.22). Maximum rise and minimum going are 220 mm. However, because the going is measured between same-handed treads (i.e. every other step), the maximum rise is, in effect, 440 mm. This gives a pitch slightly in excess of 63°. The advantage with a stair of this sort is its small footprint (Fig. 7.21e). The disadvantage is that safe use is dependent both on familiarity and agility.

Alternating tread stairs must be provided with handrails on both sides and are limited to straight flights or a sequence of independent straight flights. Such a stair may serve only one habitable room together with a bathroom and/or a WC, in which case the lavatory must not be the only one in the dwelling. Note that an alternating tread stair may only be used where it is not possible to accommodate a conventional stair.

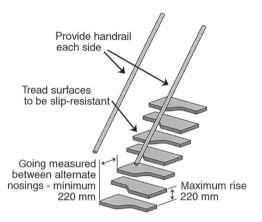

**Fig. 7.22**   Alternating tread stair.

*Fixed ladders*

Guidance on fixed ladders is similar to that concerning alternating tread stairs: a fixed ladder may serve only one habitable room and may only be used where provision of a conventional stair would require alteration to the existing space. Fixed handrails must be provided on both sides of the ladder. Note that a retractable ladder is not acceptable as a means of escape.

## Structural implications

Support for stairs is provided in a number of ways, and the structural implications must be considered at an early stage in the design process. In a conventional upper-storey straight flight (open on one side), the outer string is fixed between newel posts on adjacent floors, and the newel posts, in turn, may be fixed to floor joists or trimmers. The wall string hooks over the landing trimmer, and is either screwed or nailed to the wall.

Additional types of support are required in more complex configurations. Stairs incorporating quarter/half landings, winders (tapered treads) and extra flights, for example, may require the inclusion of load-bearing newels or other means of support at points where the stair changes direction.

As noted earlier, it is desirable in all cases for the stair to enter the conversion parallel to floor joists in order to minimise trimming.

## Stair provision: practical aspects

Providing stair access is potentially the most problematic aspect of loft conversion. Most difficulties are caused by inaccurate measurements or by a failure to properly assess the relative position of other new elements (such as modified roof slopes) and existing elements (such as the position of doorways that are 'crossed' by the stair). A number of considerations are outlined below.

### *Floor-to-floor measurement*

Because stairs must have equal risers, it is essential that an accurate floor-to-floor measurement (i.e. the overall rise of the stair) is provided (Fig. 7.20). The safest approach is to fix the conversion's floor joists and trimmers first before making the floor-to-floor measurement: the thickness of boards used for flooring in the conversion (typically 22 mm) must also be included in the measurement. An alternative – and riskier – approach is to ensure that the conversion floor is constructed to a predetermined height relative to the lower floor.

### *Headroom*

As noted above, ensuring adequate headroom is of critical importance. Downward projecting elements such as beams are potentially problematic. In addition to this, availability of stair headroom beneath a roof slope is sometimes overestimated in the initial survey, particularly if the final thickness of insulation and plasterboard is not properly taken into account. A typical roof slope may require an additional depth of structure of around 100 mm (measured at right angles to the rafters) to conform to guidance in Approved Document L. In a roof with a 35° pitch, the effect of this would be to reduce headroom, vertically, by approximately 120 mm. Where the headroom infringement covers a relatively small area, it is sometimes possible to create an increased clearance by installing a roof window. In order to reduce the risk of injury, it is of course essential that such a window be hinged at its top rather than the centre.

# 8 Wall structure

This chapter examines approaches to wall design in loft conversions. Both timber and masonry structures are considered (Fig. 8.1). Construction details, where indicated, represent generally accepted practice, although a variety of approaches may be adopted. In all cases, it is incumbent upon the designer or person carrying out the work to demonstrate, by calculation where necessary, that a proposed approach is suitable for a specific set of circumstances.

## EXTERNAL STUD WALLS

New external dormer cheek and face walls in conversions are generally constructed from timber studs and sheathing. In this section, the expression 'stud wall' or 'studwork' rather than 'timber frame' is used to differentiate loft wall construction from the conventional version of timber frame used in volume house building.

Although similar in some respects to factory-produced timber frame construction, the version of timber frame used in loft conversions differs in a number of important ways:

- External stud walls for conversions are generally built in situ, with timber components cut on site and assembled by hand.
- Tiles, slates or other vertical cladding materials – rather than brickwork – are used to provide a weather-resisting finish to the dormer face and cheeks in most cases. External walls in conversions are therefore generally thinner than conventionally built equivalents, although their thermal performance must be the same.

There are a number of advantages in adopting stud wall construction. The materials used are relatively inexpensive and sourcing them is straightforward. Stud frames are constructed swiftly and offer instant load-bearing potential: in the case of box dormer construction, this means that it is possible to provide a weather-resistant flat roof at a relatively early stage in the construction process. It is also a simple matter to accommodate thermal insulation material between the vertical studs to satisfy current guidance (see Chapter 10). In addition, the relatively low mass of stud walls is advantageous in older buildings where the strength of existing masonry walls and their foundations is not known.

### Stud arrangement and spacing

A grid approach is sometimes adopted in setting out studwork for loft conversions, and this is based on established timber engineering practice with vertical studs at 400 or 600 mm

*Loft Conversions*, Second Edition. John Coutts.
© 2013 John Coutts. Published 2013 by Blackwell Publishing Ltd.

(a)

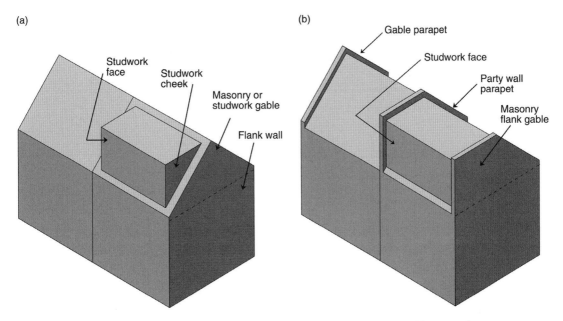

**Fig. 8.1**   Basic external wall configuration. (a) Box dormer conversion, (b) Full-width conversion.

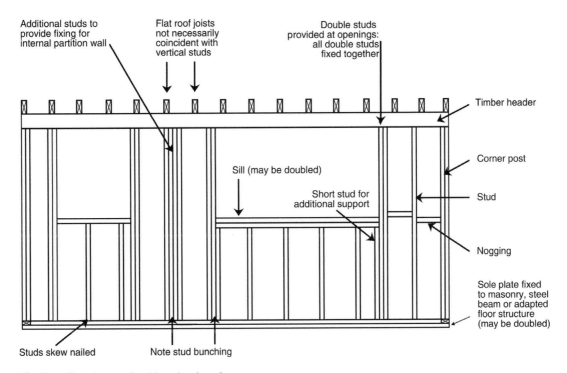

**Fig. 8.2**   Box dormer: load-bearing face frame.

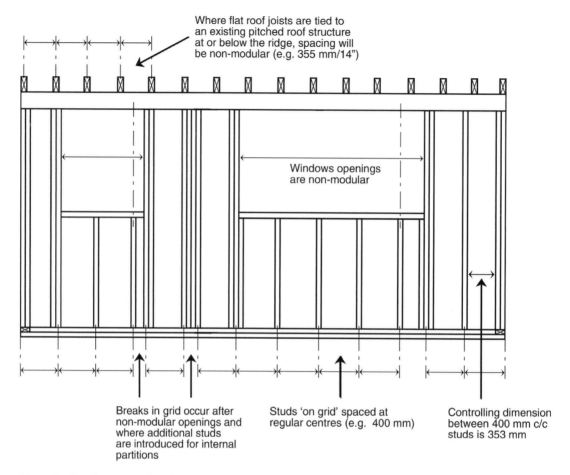

Where flat roof joists are tied to an existing pitched roof structure at or below the ridge, spacing will be non-modular (e.g. 355 mm/14")

Windows openings are non-modular

Breaks in grid occur after non-modular openings and where additional studs are introduced for internal partitions

Studs 'on grid' spaced at regular centres (e.g. 400 mm)

Controlling dimension between 400 mm c/c studs is 353 mm

**Fig. 8.3**   Box dormer: modularity.

**Table 8.1**   Dimensions of common components used in external stud wall construction.

| Component | Dimensions (mm) |
|---|---|
| Sheathing boards | **2400 × 1200**, 2440 × 1220, 2500 × 1220, 2697 × 1197, 3050 × 1220, 3050 × 1525 |
| Rigid insulation boards | **1200 × 2400** |
| Plasterboard | **2400 × 1200** |
| Windows (common nominal widths) | 300, 488, 630, 915, 1200, 1342, 1770, 2339 |
| Slate (widths) | 200, 250, 300 |
| Plain tile (width) | 165 |

Figures in bold indicate component sizes that coordinate with 400 and/or 600 mm stud spacing. The modularity of cladding (tiles or slates) is generally not considered unless there are a number of closely spaced openings with narrow tile hanging between them. Note that plain tiles (linear cover 165 mm) are the most commonly used cladding material in loft conversions.

(maximum) centres (Figs 8.2 and 8.3). This has the advantage of minimising the wasteful cutting of sheet materials where these are supplied in 1200 and 2400 mm edge lengths (but see Table 8.1). In addition, it simplifies construction by providing regular and predictable fixing positions. In loft conversions, it is common practice to use 400 mm spacing.

However, the positioning of windows and the need to provide double flanking studs for them means that it is not always feasible to rigorously maintain equal spacing across the entire width of the dormer face and in practice a rather more flexible approach must be adopted. Note also that an axial grid of 400 mm produces a controlling dimension of 353 mm between the faces of 47 mm studs and this inevitably leads to a degree of wastage when insulating material is fixed between them. The following points should also be noted:

■ Off-the-shelf windows generally do not coordinate with a 400 or 600 mm grid. This, and the need to provide double flanking studs, pushes the studwork out of grid and leads to bunching and therefore increased thermal bridging. In this sense, there is a degree of conflict between the requirement for structural stability and the need to provide insulation. Where bunching is excessive, it may be necessary to consider providing additional insulation on the inner face of the studwork (see Chapter 10).
■ Fenestration was, and is, generally based on the coordinating sizes of relatively small masonry units (typically, brick and brick fraction combinations). However, a structural grid based on 400 or 600 mm spacing offers far less subtle arrangements if it is rigidly imposed.
■ Closer stud spacing (400 mm rather than 600 mm) is preferable, particularly where heavier vertical cladding materials are to be used such as concrete or clay tiles.
■ It is necessary to form a positive nailed connection between the sheathing and the studwork. Corner connections require careful attention. Three corner-post configurations are illustrated in Fig. 8.4.

## Elements of stud wall construction

Timber studwork loft walls are generally constructed from C16 or C24 softwood. Sawn sections may be used, although machined sections are now generally chosen instead because these provide a regular surface for fixing external sheathing and internal sheet materials such as insulation and plasterboard. Timber studs are cut square, butt jointed and nailed together (Fig. 8.5). Framing anchors may also be used in this application.

In the example shown in Fig. 8.2, dormer face studs are skew (slant) nailed at the base to sole plates and fixed at their upper extremities to the header. Note that in most box dormer configurations, the face studs support the header and must therefore be capable of supporting a flat roof load.

Studwork cheek walls are generally not configured for roof loading (Fig. 8.6). However, where built directly off a structural floor, they may be used to support the edges of the trimmed opening in the original roof slope. Cheek studs are fixed to a single head plate at the top (Fig. 8.4a) or run to full roof height and fixed to the flat roof joists (Fig. 8.4b).

## Terminology

There is a degree of regional and local variation in the terms used to describe building elements. There are few cut and dried definitions. Where a number of different terms are used to describe a given component, the most commonly accepted expression is used. Variant names are also provided.

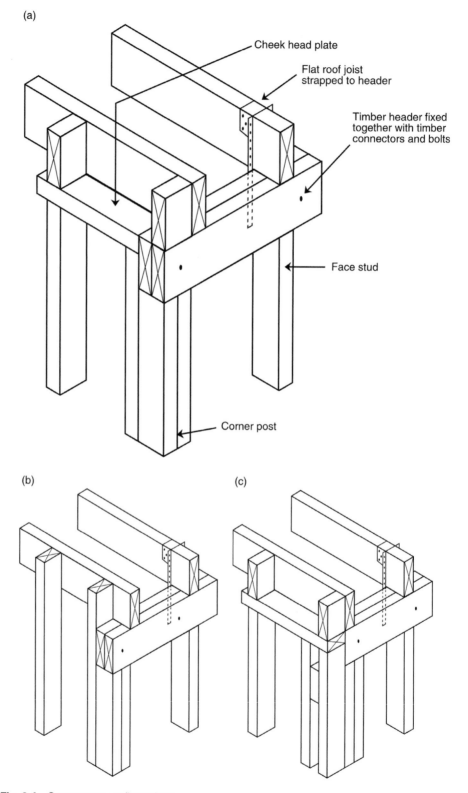

**Fig. 8.4** Corner post configurations.

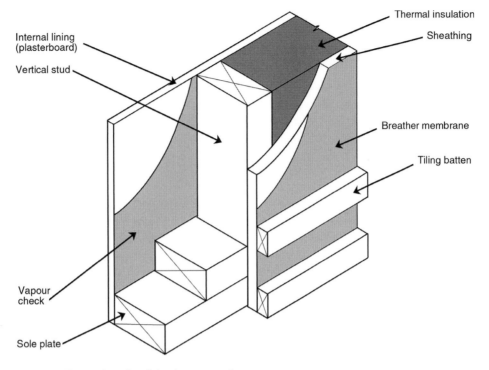

**Fig. 8.5**  External stud wall: basic construction.

Internal lining (plasterboard)

Vertical stud

Vapour check

Sole plate

Thermal insulation

Sheathing

Breather membrane

Tiling batten

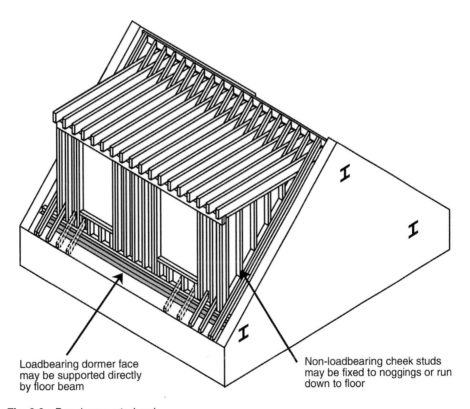

**Fig. 8.6**  Box dormer studwork.

Loadbearing dormer face may be supported directly by floor beam

Non-loadbearing cheek studs may be fixed to noggings or run down to floor

### Sole plates

Sole plates (sometimes also called plates, sill plates, floor plates, base plates, sills/cills or bottom rails) provide a fixing for the feet of vertical studs (Fig. 8.2). The sole plates themselves may be fixed to existing masonry, to beams or to a new structural floor. Effective anchoring of the sole plate must be considered in relation to wind loads, including the effects of wind uplift on the relatively low-mass structure. Both single and double sole plates may be used depending on loading conditions and the structural elements to which they are fixed.

Sole plates may be fixed in a number of ways depending on the design of the conversion. These include nailing or heavy duty screwing to joists in the new floor structure, bolting to steel floor beams, shot-firing to steel beam flanges or fixing directly to existing masonry with straps, expansion fittings or chemical anchors (Fig. 6.15). The use of expansion fittings requires particular care with older and often friable brick masonry, and strapping or chemical anchoring should be considered.

### Studs

Softwood studs with minimum dimensions of $100 \times 47\,mm$ ($97 \times 47\,mm$) are set at 400 or 600 mm (maximum) centres to support anticipated loads and to conform to the dimensions of sheet materials. Where they are to be load-bearing, wall studs are designed as compressive members with a minimum strength class of C16. Studs are skew nailed with at least two nails for each joint. The choice of external cladding may also influence stud spacing, particularly where heavier materials are to be used. Tiling battens must be nailed through to the studs rather than to the sheathing alone. Structural studs should not be cut into to provide a housing for noggings, sills, lintels or other horizontal members: where additional support for these is required, short studs or cripple studs should be used instead.

### Noggings

Fixing one row of staggered noggings at half stud height is recommended in load-bearing walls (Fig. 8.2). Noggings that have the same sectional dimensions as studs are introduced where there is an increased need for rigidity and to resist buckling. They are also used to provide support for heavy internal fittings such as radiators and sanitary appliances. Noggings may also be provided at the sheet edges of wallboards and sheathing to provide fixing positions. They are sometimes described as noggins, nogs or dwangs.

### Header

The header is the uppermost horizontal element of a load-bearing dormer face frame in a box dormer conversion and it generally provides a bearing for flat roof joists (Figs 8.2 and 8.4). This component is variously described as a head beam, header beam, window header, top head, header joist, head plate or double head-plate beam.

The header is generally formed from a pair of deep-section joists but, as with other elements of horizontal load-bearing structure, it must be designed by a structural engineer: most local authorities require that calculations be submitted. In box dormer conversions, however, it should be noted that the header seldom spans clear at full length, and face studs generally provide intermediate support.

The advantage with using a header which has a considerably deeper section than the thinner rail/binder combination used in mainstream timber frame construction is that it provides a ready-made and continuous lintel. It is also robust enough to support a new flat roof structure and at the same time allows a degree of offsetting of point loads. This frees both the designer and the builder from the need to meticulously line up each flat roof joist with a vertical stud. Equally, it offers a degree of freedom in positioning the vertical studs, provided they are not more than 400 mm apart. The ability to make such changes contingent upon conditions that arise during a conversion considerably speeds up the construction process.

### External sheathing

Sheathing boards (sometimes called bracing boards) are usually exterior grade plywood or oriented strand boards (grades OSB3 or OSB4). These are nailed to the external faces of the studs. In conventional timber frame construction, sheathing boards may be as little as 8 mm in thickness; in loft conversions, 12 or 15 mm boards are generally used.

The primary function of sheathing is to stiffen the wall and provide resistance to racking deformation. Sheathing also assists in reducing wind penetration and provides a supporting background for both breather membrane and insulation materials.

Nail specifications and nailing schedules are influenced by the thickness of the sheathing material and loading considerations. However, with plywood or OSB up to 12 mm thick, the use of 3 mm diameter corrosion-resistant 50 mm nails would be reasonable. Boards are nailed to studs and horizontal members at 150 mm centres along their edges and at 300 mm intervals mid-panel. A gap of 3 mm is allowed between adjacent boards (Fig. 8.7).

Manufactured board sizes vary depending on the country of origin. A proportion of plywood used in the UK is manufactured in North America and in many cases delivered boards measure 2440 × 1220 mm (i.e. 8' × 4') and will therefore require cutting to conform to a metric structural grid. Equally, stud positions can be adjusted to accommodate the additional length, but note that internal wallboards conform to metric dimensions (1200 × 2400 mm).

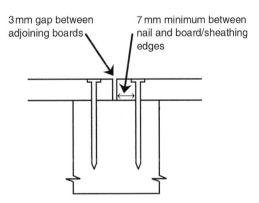

**Fig. 8.7**   External sheathing: fixing to studwork.

### *Breather membrane*

It is practice to cloak the external face of the sheathing with breather membrane. There are a number of proprietary versions but all of them work in a similar way, allowing water vapour to escape from the structure while preventing any wind-driven rain that has penetrated the tile/slate cladding from entering it. In addition, breather membrane is an effective barrier to wind penetration.

Breather membrane is fixed to sheathing with stainless steel staples. A 100 mm lap should be made at horizontal joints with the upper sheet always overlapping the lower. At vertical joints, the lap should be a minimum of 150 mm.

The use of roofing underlay or felt rather than breather membrane is generally considered to be unsatisfactory. Traditional roofing underlay has low vapour permeability and therefore increases the risk of interstitial condensation either on the sheathing or within the wall structure. Guidance provided in Approved Document C *Site preparation and resistance to contaminants and moisture* indicates that cladding should be separated from insulation or sheathing by a vented and drained cavity, with a membrane that is 'vapour open'.

## Openings

Where openings for windows are to be formed, additional flanking studs should be provided at each side (Fig. 8.2). It is practice to provide a sill or double sill beneath window openings. Lintels, where used, are supported by cripple studs. In most small to medium-sized conversions, however, the header acts as a lintel.

## Supporting structural steel in stud walls

Appropriately designed, timber frame walls are capable of providing support for structural steel components such as roof beams (Fig. 8.8a). The studs, sheathing and supporting structure below must be correctly adapted if the wall is to fulfil this function. Generally, a timber post is built into the stud wall and this may be either a solid timber section (Fig. 8.8b) or a number of studs fixed together. Structural engineering calculations are required to prove the adequacy of the beam, the timber post and the structure that will support it.

## VERTICAL CLADDING

Treated timber battens for vertical tile or slate hanging are fixed on top of breather membrane. The gauge (vertical spacing between horizontal battens) is primarily governed by the choice of cladding material (Table 8.2). The positions of window openings and other fixed points such as the junction at the bottom of the wall and the soffit/fascia at the top must also be taken into account (Figs 8.9a and b).

In all cases, battens should be at least 1200 mm in length and fixed at a minimum of three supporting points. Butt joints should be staggered and splay-cut joints cut at 45°. Care should be taken to ensure that battens are nailed through the sheathing and directly

(a)                                         (b)

**Fig. 8.8**   Roof beam support over stud wall opening. (a) Lintel and vertical studwork, (b) Timber post in stud wall.

**Table 8.2**   Vertical cladding.

| Cladding material | Unit dimensions (mm) | Batten size (mm) | Typical gauge (mm) |
|---|---|---|---|
| Plain tiles | 265 × 165 | 38 × 25 | 114 |
| Slate (ladies) | 400 × 200 | 50 × 25 | 155* |
| Slate (countesses) | 500 × 250 | 50 × 25 | 205* |

*65 mm head lap assumed.

into the studs – nailing to sheathing alone would generally not be an acceptable means of support. Where battens are cut, they should be sawn cleanly and the ends treated with timber preservative.

In exposed locations where there is an increased likelihood of wind-driven rain penetrating joints in the cladding, vertical counter battens should be fixed behind the horizontal ones to create a void from which any water may drain away freely and within which air may circulate. Approved Document C *Site preparation and resistance to contaminants and moisture* (2004) indicates that where cladding is supported by timber components, the space between the cladding and the building should be ventilated, although the use of counter battens is not specifically referenced.

Note also that the Approved Document specifically recommends the use of 25 mm checked rebates for window and door reveals in areas of high exposure to driving rain.

(a)                                              (b)

**Fig. 8.9**   Vertical cladding at junctions. (a) Tilehanging at base of cheek, (b) Slates and soaker at wall return.

## FIRE RESISTANCE OF DORMER STUD WALLS

While a normally constructed solid masonry wall (225 mm) or a cavity masonry wall (brick 100 mm, cavity 50 mm, block 100 mm) is generally assumed to exceed fire resistance requirements for the purposes of a loft conversion, the same assumption cannot be made about an external timber stud wall.

Fire resistance and space separation are examined in detail in Approved Document B *Fire safety*. In simple terms, however, where a stud cheek wall is within 1000 mm of a relevant boundary it is generally considered to constitute an 'unprotected' area and this is generally the case irrespective of the vertical cladding material used. Where such a wall has an area greater than $1\,m^2$, it requires 30-minute fire resistance to both sides (Fig. 8.10).

The following measures are generally taken to satisfy the guidance set out in Approved Document B:

- External lining: a fire-resistant board providing 30-minute fire resistance is fixed on top of the sheathing before breather membrane, battens and tiles are fixed. Proprietary fire-resistant boards are produced by a number of manufacturers and are generally made from calcium silicate.
- Internal lining: a generally accepted solution is to fix a 12.5 mm plasterboard lining with scrimmed joints and a gypsum plaster finish.

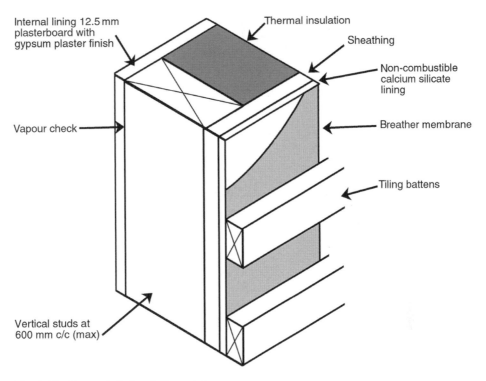

Internal lining 12.5 mm plasterboard with gypsum plaster finish

Thermal insulation

Sheathing

Non-combustible calcium silicate lining

Vapour check

Breather membrane

Tiling battens

Vertical studs at 600 mm c/c (max)

**Fig. 8.10**   External stud wall: fire resistance.

## MASONRY WALLS (EXTERNAL)

Brick masonry may be used in the construction of gable, flank and party walls in loft conversions. Note that the Party Wall etc. Act 1996 should be invoked where work on a party wall is proposed (Chapter 1).

The use of brickwork for gables and flank walls is often advantageous for a number of reasons. Building up a wall in brick similar to the original may be aesthetically preferable to tiled, slated or rendered stud walls. This is particularly the case in end-of-terrace and semi-detached conversions where exposed flanks and gables are a dominant feature.

In terraces where a number of conversions are likely to take place, the use of masonry walls between conversions introduces a degree of vertical discipline and goes some way to eliminating the often ragged and unsightly junctions that occur where stud-built conversions abut.

Brick masonry flank walls, unlike stud walls in most cases, transmit their self-load directly to the foundations. This may simplify floor design to some extent. In addition, masonry flanks provide a potential path for some flat roof loads that might otherwise require support from the floor or floor beams (Fig. 8.11). The following points should also be noted:

- Constructing a masonry gable or flank gable wall, rather than building a floor-supported inset stud cheek, maximises the amount of useable space inside the conversion.
- Brick masonry, both cavity and solid (225 mm), offers good resistance to fire and generally requires no additional detailing to achieve this.

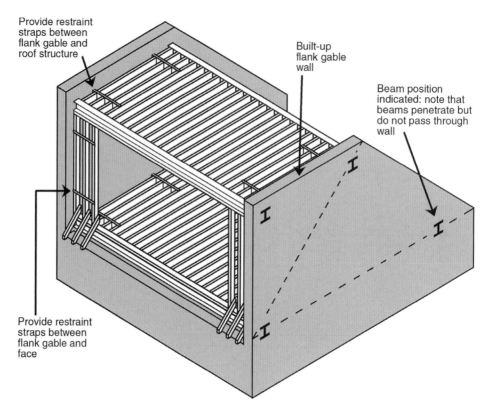

Provide restraint
straps between
flank gable and
roof structure

Built-up
flank gable
wall

Beam position
indicated: note that
beams penetrate but
do not pass through
wall

Provide restraint
straps between
flank gable and
face

**Fig. 8.11**   Full-width dormer: masonry flanks.

- The relatively high mass of brick masonry makes it an effective barrier to the transmission of sound.
- Correctly detailed, a masonry wall requires little or no maintenance during the life of the building.

Against these advantages must be weighed the relatively higher material costs of building in masonry and the fact that brick construction is generally more time consuming and therefore more costly than erecting studwork (although this does not apply to walls built in solid blockwork – see below). Equally, additional thermal insulation would normally be needed where solid brick masonry walls or traditional narrow (50 mm) cavity walls are to be built (see Chapter 10) and this would normally need to be applied to the interior face of the wall.

Consideration should also be given to the physical condition of existing party/external walls, and the possible structural consequences of increasing the masonry loading on them, before undertaking work.

## Hip-to-gable conversion

Hip-to-gable conversions – which involve stripping out the original pitched roof hip – may be carried out in dwellings with both solid and cavity masonry walls. The new gable can be constructed in either studwork or masonry.

A variant approach, which may equally be applied to dwellings that already have gable walls, is to extend the gable wall rearward to form a flank wall. Where the dwelling to be converted has a party wall parapet, that is, a parapet that is higher than the roof covering and ridge, the extended flank wall will also be higher than the level of the flat roof (Fig. 8.1b). It must, therefore, be detailed with an appropriate coping and flashings to the new flat roof (see also *Parapet walls*).

## Safety considerations during construction

It is stressed that the process of constructing gable and flank walls is uniquely hazardous in loft conversions. By their very nature, gable walls do not enjoy the same degree of resistance to lateral loading as ordinary walls, which are buttressed by returns at each end. The same also applies to flank gables in many cases. During construction, such free-standing walls are highly vulnerable to wind loading and accidental impact, with the consequent risk of collapse.

In new build, it is practice to erect roof trusses and bracing structure *before* gables are constructed (see Fig. 9.24). The gable is progressively strapped to the roof structure as the masonry is brought up, and a margin of safety is thus maintained. This approach, of course, is rarely possible in a loft conversion. For this reason, the use of shoring and strutting must be considered, to ensure the stability of the new brickwork during the course of construction. This should remain in place until the new wall is strapped to bracing roof and wall elements.

## Lateral restraint of flank gable walls

Any design that incorporates new masonry flank gable walls must take account of the need to provide adequate lateral restraint for them. While Approved Document A *Structure* does not provide guidance specific to loft conversions with masonry walls and flat roofs, it provides general guidance on buttressing for walls and lateral support for gables.

In the case of flank gable walls, a feature of most designs is that the wall is generally not returned at its end, and thus it is not buttressed in the normal way. Other positive methods of restraint for such a wall must therefore be provided (Fig. 8.11). The form of such restraints is dependent on the design of the conversion, and any system of restraints must be designed and specified by a structural engineer or other competent person. The following elements may be considered:

- *Floor level*: restraint provided by tension straps from wall to floor structure (see Chapter 7)
- *Wall*: provision of tension straps tying the masonry flank to face studwork
- *Roof level*: provision of tension straps tying the head of flank wall to flat roof joists

In most cases, flank gable walls are raised in solid masonry. Approved Document A does not provide specific guidance on fixing tension straps to solid masonry walls, but Fig. 7.15b illustrates one possible method of achieving this.

## Brick selection and size

Before undertaking any conversion where new brickwork is to be used in conjunction with existing walls, careful measurements of the existing bricks and brickwork should be made and a reliable source of matching bricks, either new or reclaimed, identified.

It should be noted that bricks with standard dimensions did not appear until 1904 when the Royal Institute of British Architects (RIBA) adopted the Southern Brick Standard with a brick height of 2 5/8" (66.8 mm). Until 1904, there were only popularly used sizes of bricks and even these were subject to considerable local and regional variations.

Other standards were introduced during the twentieth century. RIBA introduced an additional standard – the 'northern' brick standard – of 2 7/8" (73 mm) in the 1920s. As their names suggest, the 'northern' and 'southern' standards reflected regional usage. However, these bricks were standard only in name, and local variations continued until the introduction of the imperial standard brick in 1965 with measurements of 8 5/8" × 4 1/8" × 2 5/8" (219 × 104.8 × 66.8 mm).

The current British Standard metric brick was introduced in 1974 and is 215 × 102.5 × 65 mm. However, some imperial-dimensioned bricks are still manufactured (see below).

### Reclaimed bricks

There is a degree of cachet associated with the use of reclaimed bricks and these are now used extensively in hip-to-gable conversions. Local or stock bricks are routinely recovered during demolition work, and in many areas these provide a source of well-matched bricks in terms of colour, composition and size. Bricks are cleaned and palletised for reuse and are generally supplied in batches of 500. Their relatively high cost – yellow stocks, for example, are generally two-and-a-half times the price of common bricks – reflects the labour-intensive nature of the recovery process.

It should be noted that the physical characteristics of individual bricks in reclaimed batches may vary considerably and it is unlikely that a supplier will be able to provide details of, for example, the compressive strength and frost resistance of the bricks. The latter is of importance where the wall is to be carried above the roof decking as a parapet. The potentially limited crushing strength of such bricks is an additional consideration, particularly where the wall to be constructed is intended to support a beam. Careful attention should be given to the specification of padstones or bearing plates in such cases (see also Chapter 6).

Laying reclaimed stock bricks in a satisfactory manner is an important additional consideration. Dimensional irregularities are a feature of such bricks, and achieving a visually acceptable result is a test of the bricklayer's craft. The use of cement:lime:sand mortar, rather than cement:sand, is recommended for reclaimed bricks. With softer stocks, consideration should also be given to the use of a designation iv (weaker) mortar to reduce the risk of the brickwork cracking.

### New bricks

There are more than a thousand different types of bricks available in the UK and it is often possible to match, both in appearance and size, new bricks with old. As well as their lower cost, there are a number of advantages in specifying new rather than reclaimed bricks. Their characteristics are declared (e.g. crushing strength, frost resistance and the presence of soluble salts). There is also generally little wastage due to selection.

Many brick manufacturers make and hold bricks in various imperial sizes. The most common imperial-compatible new bricks are (in height): 80 mm (3 1/8"), 73 mm (2 7/8"), 67 mm (2 5/8") and 50 mm (2"). The majority of these bricks are 215 mm long and 102.5 mm wide.

It is rarely possible to achieve a perfect tonal and textural match with any new brickwork and a degree of visual discontinuity is inevitable, even if it is only the jointing that stands out. It is possible to harmonise the appearance of mismatched brickwork by tinting. British Standards exist for the pigments used in tinting, but there are currently no standards specifying tinting procedures.

Where suitable new or reclaimed stocks are not available, bricks to match may be made to order. A number of brick manufacturers are able to do this, although for small batches, less than 1200 for a typical solid masonry hip-to-gable conversion, this may not be considered economical.

## Solid blockwork

Gables or flank gables may also be built up with lightweight aerated concrete blocks that are the same width as the wall (Fig. 8.12). This approach may be adopted in buildings with either existing solid or cavity walls. Lightweight blockwork has the advantage of offering a relatively high level of thermal insulation, a relatively low mass and, because individual units are comparatively large (standard face dimensions of blocks are 440×215 mm), walls may be constructed relatively swiftly. Note that a blockwork wall needs a final weather-resisting finish of render or cladding and will also require supplementary internal insulation to meet current guidance (see Chapter 10).

Aerated concrete blocks are produced in a range of sizes that coordinate reasonably well with most existing walls, cavity or solid, and blocks are generally available in 215 and 265 mm thicknesses. Because they have a relatively low compressive strength, they should be laid with one of the weaker mortars: designation iii is generally used in this application. In the interests of achieving a better bond, some manufacturers recommend that blocks be laid in cement:lime:sand mortar. Blocks should be laid with a regular bond pattern with a minimum overlap of a quarter of a block. Mortar joints, both inside and out, should be left recessed to provide a key for render and plaster finishes.

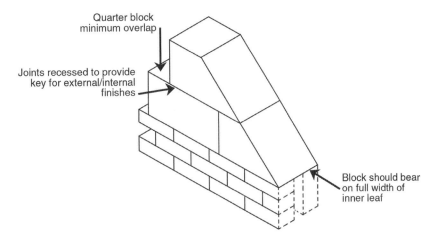

**Fig. 8.12**   Full-width aerated concrete blockwork for new gable wall.

Note that while standard 440×215×100 mm blocks can be laid face down to produce a 215 mm wall, this approach would result in an undesirable increase in thermal bridging. It is also stressed that new blockwork should be bedded across the full width of the existing inner (load-bearing) leaf in a building with a cavity wall.

## Mortar and brickwork

In low-rise buildings, mortars of designations iii and iv are generally used, depending on brick type and exposure. Note, however, that Approved Document A contains only guidance related to mortars of designation iii or stronger.

In cement:lime:sand mixes, ordinary Portland cement and dry hydrated lime (calcium hydroxide) are generally used. Hydrated lime improves workability, provides a better bond and offers enhanced water resistance. Sand is now generally described as fine aggregate in guidance. Where cement:sand mortar is used, it is generally mixed with an air-entraining agent (plasticiser) to improve workability.

The higher the proportion of cement (binder) in a mortar mix, the higher its compressive strength will be. However, because strong mortars are less flexible, they are more likely to cause destructive localised stresses. For this reason, mortar should generally have a lower compressive strength than the bricks it is designed to bond.

This is of particular importance when using reclaimed bricks of unknown crushing strength. It should also be noted that the strength of brickwork is determined by both bricks and the choice of mortar: it is generally the lower strength of mortar, rather than the higher strength of bricks, that determines bearing capacity.

Where frogged bricks are to be used, it is good practice to lay them frog up. This reduces localised stresses within the brickwork, particularly on brick edges if the wall is designed to support structural elements such as beams. The wall will also have better sound insulation properties because there are less likely to be voids in the brickwork.

It should also be noted that using imperial brick sizes in conjunction with metric blockwork, as is occasionally the case with hip-to-gable conversions in inter-war

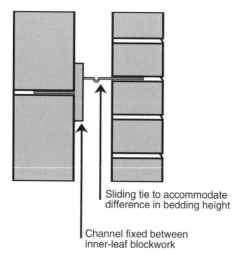

Sliding tie to accommodate
difference in bedding height

Channel fixed between
inner-leaf blockwork

**Fig. 8.13**  Two-part wall tie.

houses, will lead to misaligned bed joints caused by the relatively greater thickness of pre-metric bricks. When tying the inner and outer leaves, therefore, it is practice to use a two-part wall tie with a vertical channel section and sliding tie to accommodate differences in bedding height (Fig. 8.13). An alternative is to use helical screws. Conventional wall ties should not be bent to accommodate differences in bedding height. Note that Approved Document A now specifies the use of stainless steel ties in all domestic dwellings.

## Parapet walls in loft conversions

Rather than terminating masonry flank walls beneath the roof covering, the walls may project above the roof surface to form a parapet. Guidance on maximum heights and minimum thicknesses of parapet walls is provided in Approved Document A *Structure*. In loft conversions, though, achieving the minimum height of a parapet is often a significant consideration (Figs 8.14a and b).

Where there is a need to form a particularly low parapet wall relative to the finished roof deck, the requirement to form a weather-resisting junction between the roof covering and the parapet wall becomes the limiting factor. Note that the parapet height must be referenced to the highest point of the flat roof, taking firring and the thickness of roof coverings into account. Over a typical loft conversion flat roof span, the fall of the flat roof will generate a difference of about 75 mm (slightly more than the height of a single brick) between the highest and lowest points.

In most cases, the minimum parapet height that can be achieved for practical purposes relative to the highest point on a flat roof is about 225 mm (Fig. 8.15) plus the thickness of the coping or capping which could be coping stones, brick-on-edge or sheet metal dressed over boards. In all cases, the minimum upstand for the roof covering should be considered. This is generally 150 mm and is intended to provide a degree of protection against rainwater splash. It also limits the risk of damage caused by temporary inundation.

It should be noted that parapet walls are exposed to a relatively high degree of weathering, and particular care should be taken in their detailing to ensure durability. Typically, both sides of parapet wall brickwork are exposed. However, in a low-level parapet, exposure is reduced somewhat on the deck side of the wall, and brick choice is less of a critical factor than it would be in a relatively high parapet. In all cases, however, using a mortar of designation i or ii would be reasonable.

In order to protect the brickwork, the coping or capping (whether masonry or metal) should oversail the parapet wall on both sides. Note that the damp proof course (dpc) at cover flashing level is placed *above* the cover flashing.

## INTEGRATING NEW AND OLD

Junctions between existing roof structures and new walls are inevitable given that most loft conversions are subordinate structures. Angular relationships between new elements of structure and the existing roof are frequently complex, and providing an effective weatherproof interface is not always straightforward (Fig. 8.16).

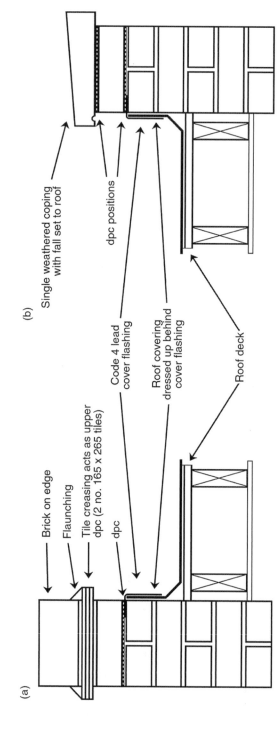

**Fig. 8.14** Low-level parapet walls in solid masonry. (a) Brick-on-edge capping, (b) Single weathered coping.

**Fig. 8.15**  Parapet treatments.

**Fig. 8.16**  Flashings and weatherings: face to slope.

Weather-resisting junctions are generally formed from lead sheet. Lead is malleable and this means that as well as being relatively easy to work, at least in simple applications, it is capable of accommodating small movements between elements of structure without fracturing. Its considerable weight, even in lighter codes ($14.97\,kg/m^2$ for Code 3 sheet, $20.41\,kg/m^2$ for Code 4 sheet), means that it is also resistant to wind uplift when appropriately clipped.

Forming junctions between two single planes (i.e. wall-wall, wall-roof slope) is relatively straightforward. Fig. 8.17 illustrates a side abutment with lead step flashing between a chimney stack and a new dormer cheek before tilehanging. Note that the brick joints are raked clean with all mortar residue removed. Each step of the flashing is folded over by

**Fig. 8.17**   Cheek to stack.

**Fig. 8.18**   Cheek to slope.

25 mm for fixing into the joint. The lead flashing is fixed in position with a lead wedge for each step before the joints are filled with mortar or a proprietary lead sealant.

Sloping junctions or abutments between an existing roof and a new tiled dormer cheek are weatherproofed using soakers, usually in Code 3 or Code 4 lead sheet. Soakers are interleaved between the tiles or slates of the existing roof slope and turned up behind the cheek cladding (see also Fig. 9.14).

A horizontal junction between a tile-hung dormer face and an existing portion of roof slope or wall is generally provided by a cover flashing that is dressed over the lower tiles or top of the wall and fixed behind the cladding of the vertical dormer face. The cover flashing should be in at least Code 4 lead, and it should be clipped to resist uplift. The use of handed 90° corner tiles provides a weatherproof junction between the dormer face and cheek (Fig. 8.18).

Fig. 8.18 also illustrates the sort of multi-plane junction that is common in loft conversions. In this case, the dormer face and dormer cheek intersect with the original roof slope. In order to adequately weatherproof a junction of this sort, lead welding (lead burning) or bossing may be required to create an appropriately shaped flashing.

Similar junctions may occur where parapet walls form a junction with other roof elements. Fig. 8.15 illustrates a junction between two parapet walls, one with coping, one capped with lead sheet, and the kerb-upstand of an asphalt roof. While detailing junctions of this sort is a relatively time-consuming process, it should be noted that most roof failures occur at junctions between planes and not within the planes themselves.

## Chimney cowls

The significance of water ingress via chimneys is sometimes overlooked. This is, in part, because flue offsets in the roof space intercept the bulk of the precipitation – rain and snow – which is then absorbed by brickwork, unnoticed, until the loft is occupied. Given that a typical four-pot stack is capable of directly admitting in excess of 30 gallons of water each year, consideration should be given to the provision of appropriate cowlings.

## COMPARTMENT (PARTY) WALLS

The wall separating adjacent but separately occupied properties in a terrace or semi-detached dwelling is generally referred to as a party wall (the term has a distinct legal meaning as well – see Chapter 1). For the purposes of fire safety, such a wall is described as a compartment wall.

Until the late nineteenth century, adjacent terraced dwellings were not routinely separated from each other by compartment walls in the roof void: walls were sometimes built up only to the level of the upper-storey ceilings. In cases where the wall is absent, separation may be achieved by raising a new compartment wall, at full thickness, in the loft space. A 215 mm masonry wall will generally conform to guidance on fire resistance. The following points should also be noted:

- Where a new party wall is to be constructed, it will be necessary to invoke the Party Wall etc. Act 1996 and to inform the adjoining owner or owners (see Chapter 1).
- A compartment wall should be taken up to meet the underside of the roof covering or deck, with fire-stopping where necessary at the wall-roof junction to maintain continuity of fire resistance. Compartmentation should also be continued across any eaves cavity.

Timber components such as beams, joists, purlins and rafters may be built into or carried through a masonry compartment wall, provided that openings for them are kept as small as possible and fire-stopped. Similarly, roofing underlay and tiling battens, if fully bedded in mortar, may be carried over a compartment wall. Guidance to this effect is provided in Approved Document B. However, the practice of 'building in' timber structural elements is generally discouraged except where no alternative exists, as in the case of tiling battens that must bridge the compartment wall in a terrace.

Mortgage lenders generally require that separation is present in the roof void and, for this reason, it may be found that separation has been provided subsequent to construction by building up the party wall in 100 mm blockwork or similar. This is unlikely to provide adequate sound insulation or a suitable bearing for new floor and roof beams, and replacement with 215 mm of solid brick masonry may be the only suitable solution.

## INTERNAL PARTITIONS

Partition walls within the conversion provide separation between rooms and may be framed in 100×47 mm (97×47 mm) timber studwork and clad with plasterboard. Where partitions form part of the enclosure of a protected stair, they must be fire resisting (see Chapter 4).

In addition, internal walls between a bedroom or a room containing a water closet, and other rooms, must provide reasonable resistance to airborne sound. This may be satisfied by adopting the detail illustrated in Fig. 8.19. Sound resistance measures need not apply if an internal wall contains a door, or if an internal wall separates an en suite WC and the associated bedroom.

## WINDOW AND DOOR SAFETY

Windows present two types of risk: the danger from falling and the risk of cutting and piercing injuries caused by broken glass. Guidance on minimising these related but distinct hazards is set out in Approved Document K *Protection from falling, collision and impact* and Approved Document N *Glazing – safety in relation to impact, opening and cleaning*.

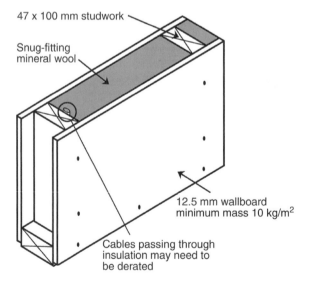

**Fig. 8.19**  Partition wall sound resistance.

## Windows

Glazed areas that are less than 800 mm above the finished floor are considered to be in a critical location for the purposes of Approved Document N because low-level glass is at a greater risk of being broken. The most widely used method of conforming to the guidance is to provide glazing that meets safe breakage criteria.

To minimise the risk of falling, the guidance in Approved Document K is that guarding be provided to opening windows at or less than 800 mm above floor level (Fig. 8.20a). However, Approved Document B *Fire safety* indicates that guarding is not required for a window in a roof slope, where the bottom of the opening may be 600 mm above the floor.

## Juliet balconies and balustrades

Inward-opening French windows have become a popular feature in loft conversions. The opening must be protected by a balustrade (often called a Juliet balcony) that is fixed either to structural studwork or to masonry (Fig. 8.20b).

The balustrade must be a minimum of 1100 mm above finished floor level and should have only a minimum of horizontal elements to prevent it being climbed. In addition, the spacing between all vertical and horizontal elements in the construction of the balustrade, including floor clearance, should be such that it would prevent a sphere of 100 mm diameter passing through the structure. A glazed balustrade, which incorporates toughened glass, may be used as an alternative. Additional guidance on guarding heights is provided in Approved Document K.

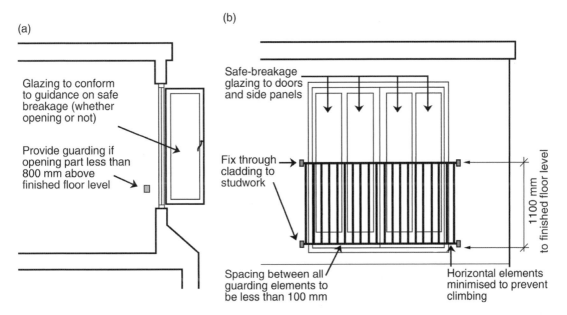

**Fig. 8.20** External openings: guarding and glazing. (a) Windows at or below 800 mm above floor level, (b) French window balustrade ('Juliet' balcony).

Normally, balustrades are bolted to masonry, but this is seldom possible in the case of a loft conversion. Fixing points capable of transmitting the necessary loadings back to the wall or roof structure should be considered at an early stage in the design process.

## Glazing requirements for doors

Doors present special hazards and the guidance for glazing in doors (such as French windows) is therefore more rigorous than that for windows: glazed door elements 1500 mm or less above floor level must conform to guidance on safe breakage. In addition, glazed side panels within 300 mm should be provided with safety glazing.

## Cleaning

The Building Regulations requirement to provide safe access for cleaning windows does not apply to dwellings. However, in order to facilitate safe and effective cleaning of external glazing from inside the property, windows of the following sort may be specified:

- Inward-opening 'tilt and turn' windows
- Sliding sash windows incorporating an inward tilting mechanism
- Casements incorporating 'easy clean' hinges

## Replacement windows

Where a window serves as a means of escape and is being replaced, the replacement unit must also meet the requirement to provide a means of escape (for dimensions, see Fig. 4.4).

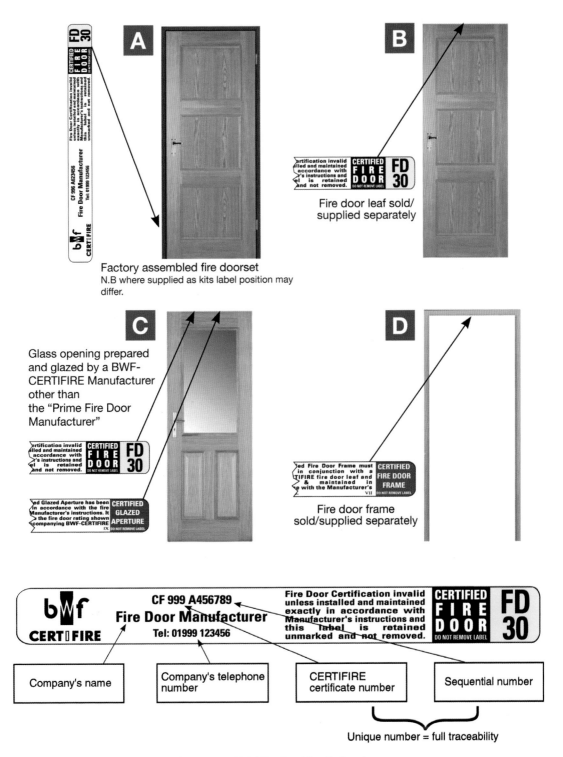

**A**

CERTIFIRE b⋁f
Fire Door Manufacturer
CF 996 A623456
Tel: 01999 123456
Fire Door Certification invalid unless installed and maintained exactly in accordance with Manufacturer's instructions and this label is retained unmarked and not removed.
CERTIFIED FIRE DOOR FD 30
DO NOT REMOVE LABEL

Factory assembled fire doorset
N.B where supplied as kits label position may differ.

**B**

ertification invalid illed and maintained accordance with r's instructions and el is retained and not removed.
CERTIFIED FIRE DOOR FD 30
DO NOT REMOVE LABEL

Fire door leaf sold/
supplied separately

**C**

Glass opening prepared and glazed by a BWF-CERTIFIRE Manufacturer other than the "Prime Fire Door Manufacturer"

ertification invalid illed and maintained accordance with r's instructions and el is retained and not removed.
CERTIFIED FIRE DOOR FD 30
DO NOT REMOVE LABEL

ed Glazed Aperture has been in accordance with the fire Manufacturer's instructions. It the fire door rating shown companying BWF-CERTIFIRE IX
CERTIFIED GLAZED APERTURE
DO NOT REMOVE LABEL

**D**

ed Fire Door Frame must in conjunction with a TIFIRE fire door leaf and maintained in e with the Manufacturer's VII
CERTIFIED FIRE DOOR FRAME
DO NOT REMOVE LABEL

Fire door frame
sold/supplied separately

bⅥf CERTIFIRE
CF 999 A456789
Fire Door Manufacturer
Tel: 01999 123456
Fire Door Certification invalid unless installed and maintained exactly in accordance with Manufacturer's instructions and this label is retained unmarked and not removed.
CERTIFIED FIRE DOOR FD 30
DO NOT REMOVE LABEL

| Company's name | Company's telephone number | CERTIFIRE certificate number | Sequential number |

Unique number = full traceability

**Fig. 4.13** Fire door certification. Courtesy British Woodworking Federation.

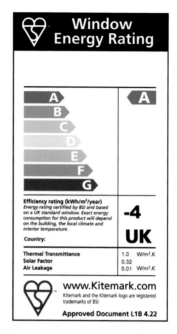

Fig. 10.10 Window Energy Rating (WER) product certification label. Courtesy BSI.

Fig. 11.2 The Zero Carbon Loft in Ealing, west London.
Courtesy Green Structures.

**Fig. 11.3** The Zero Carbon Loft incorporates a green roof, while the roof of the back addition (outrigger) has been opened up to create a terrace.
Courtesy Green Structures.

**Fig. 11.4** Thermal accumulator stores surplus heat energy.
Courtesy Green Structures.

**Fig. 11.5**  Energy-efficient LED lighting is used both inside and outside the conversion.
Courtesy Green Structures.

**Fig. 11.6**  Zero Carbon Loft roof terrace during construction.
Courtesy Green Structures.

# 9 Roof structure

This chapter considers commonly encountered roof forms and the modifications that may be made to them as part of a loft conversion. Dates provided below are for guidance only and are intended to indicate the likely use of a roofing system for a house constructed during a given period. However, roofs of any era may be hybrid structures that do not necessarily conform to textbook examples. A full understanding of the structural function of the roof must be gained before any alteration is undertaken.

## ROOF TYPES

### The cut roof (common to about 1950)

The term 'cut roof' is a fairly loose one and is used to distinguish the traditionally constructed timber roof from the trussed rafter form that has now largely superseded it.

In the cut roof, timber components such as rafters, ridge, struts, purlins and ceiling joists were sawn on site and nailed together to create a roof structure (Fig. 9.1a). This system of construction was almost universal in the UK until World War II. Generally, the structural design of such roofs was based on custom rather than calculation: it is fortunate that in many cases (but not all), cut roofs are reasonably tolerant of modification.

The sometimes generous dimensions of traditional cut roof components and the relatively small number of projections (e.g. struts and collars) within the roof void mean that these roofs are generally the easiest to convert, although a new structural floor is required in most cases.

Cut roofs often depend upon a load-bearing internal wall for some of their structural integrity. This, in itself, is often advantageous as far as loft conversions are concerned.

### The TDA roof truss (common 1947–1980)

Loft conversions in TDA timber truss roofs are generally feasible provided there is adequate headroom, but they are generally more troublesome than those in the cut roof described above. The principal difficulty lies in positioning new structural members between the struts, ties and hangers present in the roof void. However, the TDA trussed roof is rather more robust in its construction and provides more room for manoeuvre than the trussed *rafter* roof that has largely replaced it.

*Loft Conversions*, Second Edition. John Coutts.
© 2013 John Coutts. Published 2013 by Blackwell Publishing Ltd.

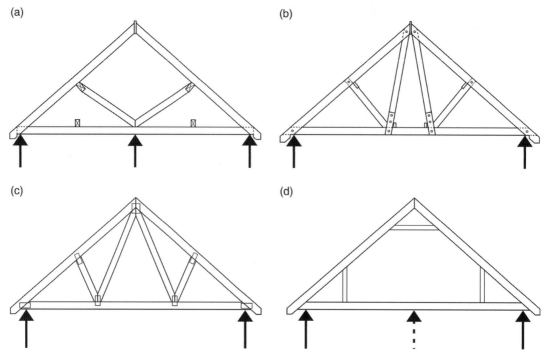

**Fig. 9.1** Domestic roofs: predominant structural forms. (a) Cut roof (common to c. 1950) generally dependent on intermediate support; rafters at 350–450 mm centres, purlin struts intermittent. (b) TDA roof truss (c. 1950–1980) may span clear between supporting walls; trusses at 1800 mm centres, infill rafters at 450 mm c/c. (c) Trussed rafter roof (c. 1965 to present) trusses generally at 600 mm centres may span clear without intermediate support. (d) Attic trusses (widespread since 2000) generally at 600 mm centres, intermediate support may be provided.

The TDA truss was developed by the Timber Development Association, now TRADA, as a response to restrictions on the use of timber that followed World War II, and it was first used in 1947. While its primary purpose was to reduce the amount of timber used in roof construction, it also allowed for the creation of greater clear spans in dwellings than had been typical before 1939. For this reason, the presence of internal load-bearing walls should not be assumed.

In a TDA truss, the primary triangular frame is created by two pairs of opposing rafters linked by a horizontal ceiling tie (Fig. 9.1b). Additional bracing within the truss is provided by struts, hangers and inclined ties. All components in the truss are fastened by bolts with split ring or toothed plate shear connectors to create a substantial and highly rigid structural frame.

The TDA truss acts as a principal, providing intermediate support for other structural members in the roof. Trusses are generally fixed at 1800 mm centres; these support purlins and ceiling binders which, in turn, support common rafters and ceiling joists set between them at 450 mm centres. In this sense, it is still very much a traditional ridge and purlin roof structure, although the design allows for purlins of considerably reduced section. A number of standard designs for bolted trusses were developed by the TDA and later TRADA.

Some elements of the TDA roof configuration are lighter in section than those of comparable components in pre-war roofs, and the spacing between them is slightly greater (450 mm rather than 400 mm for rafters and joists). Nevertheless, TDA roofs are more generously proportioned than modern trussed rafter roofs. Note that in contemporary trussed rafter configurations, spacing is generally 600 mm.

## Trussed rafter roofs (1965 to present)

There is a perception that trussed rafter roofs cannot be converted. This is incorrect. Timber trussed rafter roofs are certainly neither the easiest to convert nor the least expensive, but conversion is generally feasible provided there is adequate headroom.

In the conventional trussed rafter roof, trusses are generally fixed at 600 mm centres (Fig. 9.1c). Structural webbing and stability bracing occupy a substantial portion of the void and this makes working in the roof awkward. The major components are modestly proportioned compared to equivalent elements in cut roofs, and even minor alterations must be justified by calculation. In addition, the emergence of the trussed rafter form coincided with the vogue for much lower roof pitches and consequently reduced headroom. Note that trussed rafter roofs are often intended to span between wall plates without intermediate support. The presence of internal load-bearing walls should not be assumed.

The attic or 'room-in-roof' (RiR) truss is a variant form of the trussed rafter and is now widely used in new build (Fig. 9.1d). Webbing in the conventional sense is largely absent and a roof formed with such trusses therefore provides a useable void. Although the RiR truss emerged more than 30 years ago, it has only become popular in new build since the late 1990s. Notes on attic trusses are provided in a separate section at the end of this chapter.

## CUT ROOF: STRUCTURAL FORMS

The cut roof encompasses a considerable number of structural configurations but only common domestic forms are considered here. Generally, a distinction is made between the *single roof*, which has no purlins and the *double roof*, where purlins are present. Note that a dwelling may contain both single and double roof elements. For example, a terraced house might have a principal double roof; the back addition (see Glossary) of such a dwelling would generally have a 'lean-to' (single) roof. It should be noted that roofs do not always occur in 'pure' forms.

## Single roofs

These are constructed with rafters supported at their head and feet only: no transverse intermediate support is provided by purlins, although collars and ties are present in some of the versions described below. The span of such single roofs is relatively limited. Basic forms of the single roof include:

■ *Couple roof*: rafters are pitched from the wall plate to the ridge (Fig. 9.2a). No horizontal tying elements are provided. Maximum span on plan: 3.6 m. Not widespread as a domestic form.

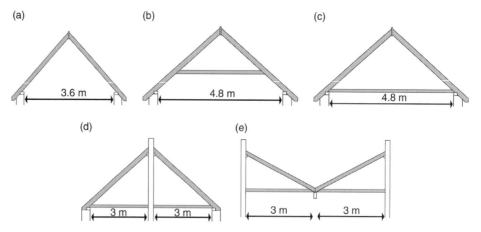

**Fig. 9.2**   Common single roof configurations (figures indicate typical maximum spans). (a) Couple roof, (b) Collar roof, (c) Close couple roof, (d) Lean-to roof, (pair), (e) Butterfly roof.

- *Collar roof*: rafters are pitched from wall plate to ridge (Fig. 9.2b). Horizontal collars are fixed between rafter pairs about one third of the way up the roof slope; rooms formed beneath the collars are thus partly within the roof void. Maximum span on plan: 4.8 m.
- *Close couple roof*: rafters are pitched from wall plate to ridge (Fig. 9.2c). Horizontal ties are provided at rafter foot level to limit the outward thrust of the roof slopes. Maximum span on plan: 4.8 m.
- *Lean-to roof*: a mono-pitch roof with rafters running from a wall plate at the foot to a wall plate or bearing plate at the top (Fig. 9.2d). This is a common method of roofing rearward subsidiary projections of terraced dwellings (back additions or outriggers), where the roof slope of the projection is at right angles to the principal roof. Ceiling joists are generally present, but may not form a true tie. Maximum span on plan: 3 m. Roofs of this sort were sometimes constructed with purlins and ceiling binders to offer a greater span.
- *Butterfly roof*: sometimes called a 'London' or V-roof (Fig. 9.2e). Rafters are pitched to opposing compartment walls from a central timber beam. Generally associated with terraced dwellings particularly in central and inner London (c. 1790–1870). Ceiling joists are present, but generally do not provide a true tie. Often characterised by an exceptionally shallow roof pitch (down to 15°). Maximum span on plan: 3 m + 3 m. As with the lean-to roof, a butterfly roof could be constructed with purlins to provide a greater span.

## Double roofs

Rafter length (clear span) is generally the limiting factor in the construction of the single roof. In a double roof, however, considerably greater spans are achieved by providing intermediate support for rafters by introducing purlins (Fig. 9.3). Double roofs are far more common in dwellings suitable for conversion, and a number of external forms are represented by this mode of construction, including gabled and hipped roofs. Equally, there are a considerable number of internal structural forms. Some of the more common configurations are illustrated in Fig. 9.4.

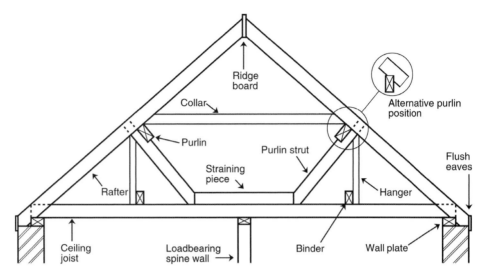

**Fig. 9.3** Common double roof elements.

## CUT ROOF: STRUCTURAL ELEMENTS

The following notes provide a description of the functions of the major components in both gabled and hipped cut roofs (Fig. 9.5). Modifications to elements of the roof structure are outlined in *Common conversion alterations* below. Note that spacing between elements, such as rafters and ceiling joists, is not always regular.

### Purlin

The purlin is generally the most substantial of the timber members in the traditional cut roof. These are fixed in a horizontal position and provide intermediate support to reduce the span and therefore the section size of the rafters. Depending on the orientation of the purlin (Fig. 9.3), rafters are fixed by skew nailing directly to the purlin face, or are nailed and notched to the purlin with a bird's mouth (or birdsmouth) cut to the rafter.

In slopes between 4 and 5 m in length, support is generally provided by a single purlin half way up the slope. To limit rafter span in dwellings with a larger plan area and longer slopes, purlins may be positioned at one third and two thirds of the way up the slopes.

In a single-purlin slope, the effect of removing the purlin is to double the span of the rafters, and, in most cases, this would not be acceptable unless an alternative means of support were to be provided. Table 9.1 provides an indication of maximum spans for rafters with dimensions of $100 \times 50$ mm (the nearest metric equivalent to the commonly used imperial 4" $\times$ 2" section).

Purlins generally require intermediate support of their own and this is provided in a number of ways depending on the design of the roof and the building. Support for purlins may be provided in the following ways.

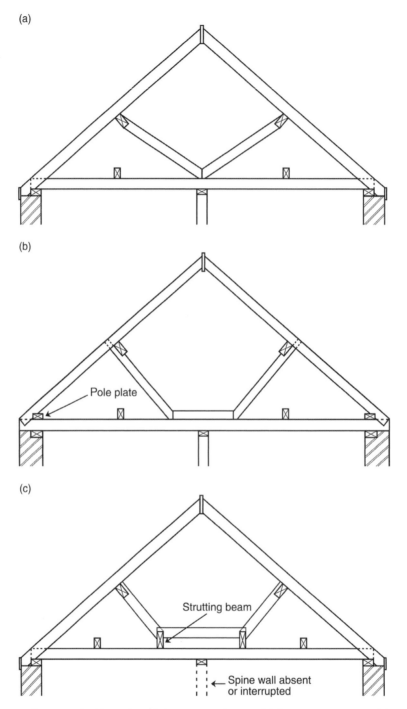

(a)

(b)

Pole plate

(c)

Strutting beam

Spine wall absent or interrupted

**Fig. 9.4**  Common double roof configurations: variations. (a) Struts to spine wall, (b) Pole plate, (c) Strutting beam.

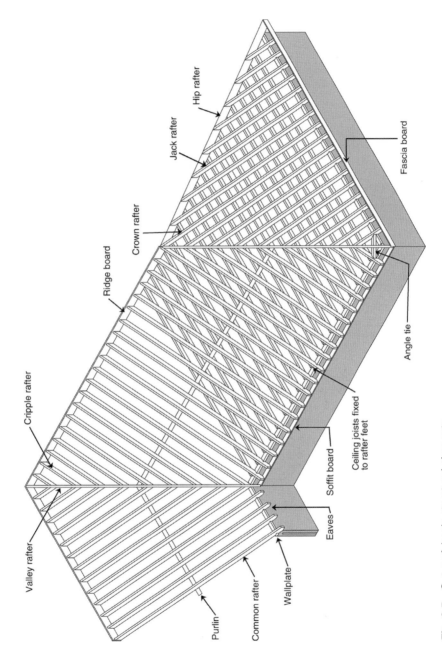

**Fig. 9.5**  Cut roof: basic structural elements.

Hip rafter

Jack rafter

Fascia board

Crown rafter

Ridge board

Angle tie

Cripple rafter

Ceiling joists fixed
to rafter feet

Valley rafter

Soffit board

Eaves

Purlin

Common rafter

Wallplate

**Table 9.1**   50 × 100 mm rafter – maximum clear spans (roof pitch 30–45°).

| Joist spacing (mm) | C16 clear span (m) | C24 clear span (m) |
| --- | --- | --- |
| 400 | 2.38 | 2.49 |
| 450 | 2.30 | 2.40 |
| 600 | 2.09 | 2.18 |

Imposed snow loading 0.75 kN/m².
Access only for maintenance or repair.
Dead load 0.75–1.25 kN/m² (excluding self-weight of rafter).
For general guidance only. Based on Approved Document A (1992).

### Gable wall support for purlin

A purlin is sometimes built into a supporting masonry gable although, for reasons of fire safety, there is a longstanding presumption against this. Support may also be provided by brick corbels projecting from a masonry gable (common, see Fig. 5.3) or by a shoe or hanger fixed to the gable. In many cases, the purlin abuts the gable wall but support is provided by struts rather than the wall itself.

### Strut support for purlin

Inclined purlin struts for opposing slopes may be positioned directly opposite each other with the load from the purlins transmitted to a central spine wall. Alternatively, in cases where a central spine wall is not present, the struts may be supported by deep binders at ceiling joist level (sometimes called strutting beams) that are suitably dimensioned for both roof and ceiling loads (Fig. 9.4c).

Typically, purlin struts are positioned at 90° to the roof slope, generally in opposing pairs. A horizontal straining piece may be introduced at the base of the struts on wider spanning roofs (Fig. 9.3). Purlin struts should not be confused with binder hangers which are generally fixed vertically near the purlin and are supported by the roof structure.

### Principal rafter support for purlin

In this configuration, intermediate support for the purlins is provided by a truss within the roof structure itself, rather than by an internal wall. Support may be provided by a traditional principal rafter truss (generally a pair of tied rafters of substantially greater section than the common rafters) or a TDA-type truss described at the beginning of this chapter (Fig. 9.1b).

## Ridge and rafters

### Ridge or ridge board

This is a horizontal timber member at the apex of the roof to which the heads of the common rafters are nailed. Typically, this is between 25 and 32 mm in breadth. The ridge is generally configured to project 25 mm above the rafter heads.

### Common rafters

These are pitched from the wall plate to the ridge and provide support for the roof covering, usually via battens. There is generally a plumb cut to the head of the rafter (rather than a bird's mouth) and the rafter is fixed by nailing to the ridge board. In order to achieve balance, rafters of opposing slopes are fixed directly opposite each other at the ridge. The foot of the rafter is generally fixed to the outer edge of the wall plate with a bird's mouth notch and skew nailing. In all cases, the bird's mouth must not exceed one third of the rafter depth. Note that the rafter feet may be bird's mouthed to the *inside* of the wall plate where eaves are sprocketed. Where a purlin provides intermediate rafter support, the rafter is nailed and is sometimes bird's mouthed over it as well. Rafters are not necessarily continuous over purlins. In older buildings, rafter spacing is not always equal across the slope.

### Valley rafter (valley structures)

This provides a fixing for cripple rafters where roof slopes meet to form an internal angle. Note that the upper edge of the valley rafter does not project above the cripple rafters.

### Hip rafter (hip structures)

The inverse of a valley rafter. It provides a common fixing for the heads of jack rafters in a hipped roof configuration. The hip rafter is considerably deeper in section than the rafters it supports. The upper edge of the hip rafter does not project above the jack rafters.

### Hip board (hip structures)

Sometimes used as an alternative expression for hip rafter (above). It also has a different meaning and is used to describe boards fixed either side of the hip rafter to provide support for slates.

### Crown rafter (hip structures)

The central rafter in a run of jack rafters in a hipped configuration. It forms a junction with both hips at the apex. Occasionally omitted.

### Jack rafters (hip structures)

These run from the wall plate to the hip rafter with a compound angle cut to the head. As with common rafters, they are skew nailed and fixed in pairs in order to maintain balance.

### Cripple rafters (valley structures)

These run from the ridge to the valley rafter. They are skew nailed in opposing pairs. Also described as jack rafters.

### Saddle board (hip structures)

A board fixed vertically to the last pair of common rafters at the end of the ridge. It provides a fixing for crown and hip rafters.

## Wall plates

These horizontal timber members are fixed to the head of the wall to provide a bearing for rafter feet and ceiling joists. In both solid masonry and cavity masonry structures, wall plates are usually set in line with the inner face of the wall. Wall plates are generally 100 × 75 mm or 100 × 50 mm. They are bedded level in mortar to provide an even bearing surface. At junctions and breaks, it is practice to form half-lapped joints.

### Strapping for wall plates and rafters

Traditionally, vertical restraint fixings were not routinely provided for wall plates and rafters, the mass of the roof structure and its tile or slate covering being considered sufficient to resist wind-generated lateral and uplift forces.

Current guidance for new roof structures is that vertical strapping at least 1 m in length should be provided at eaves level at intervals not exceeding 2 m, although the Approved Document outlines a number of exceptions to this. Note that there may be a need to retrofix vertical restraint straps to an existing pitched roof slope if the loading conditions are changed.

## Ceiling joists and collars

*Ceiling joists* provide the primary ties between opposing roof slopes in the majority of dwellings. The purpose of these ties is to limit the tendency of a pitched roof to push outwards against (spread) its supporting walls. They are fixed to wall plates and rafter feet at both sides of the building. Ceiling joists also provide support for ceilings. Because the spans involved are often considerable, each tie may comprise two ceiling joists nailed together at a central point, in effect creating a single continuous member.

*Collars* are sometimes provided in addition to ceiling joists to enhance the tie between slopes. Note that the collar roof (Fig. 9.2b) is a structural form in its own right in which the collar performs the primary tying function.

## CUT ROOF: COMMON CONVERSION ALTERATIONS

## Modification of the roof structure

In order to create a useable roof void free from structural intrusions, the system of support for the original roof slope or slopes will be substantially altered (Figs 9.6 and 9.7). The following modifications may be carried out:

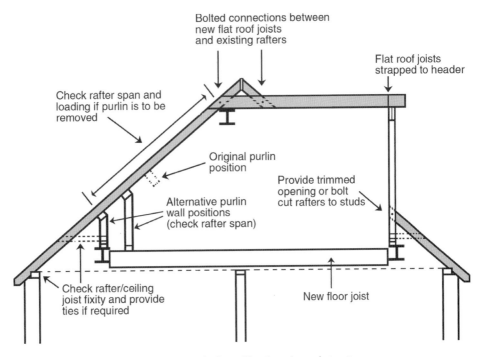

Bolted connections between new flat roof joists and existing rafters

Flat roof joists strapped to header

Check rafter span and loading if purlin is to be removed

Original purlin position

Provide trimmed opening or bolt cut rafters to studs

Alternative purlin wall positions (check rafter span)

Check rafter/ceiling joist fixity and provide ties if required

New floor joist

**Fig. 9.6** Large dormer conversion: typical modifications to roof structure.

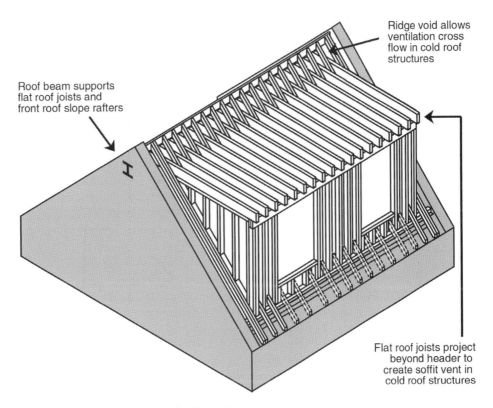

Ridge void allows ventilation cross flow in cold roof structures

Roof beam supports flat roof joists and front roof slope rafters

Flat roof joists project beyond header to create soffit vent in cold roof structures

**Fig. 9.7** Large dormer conversion: flat roof.

- Removal of original purlin(s) and struts. Replacement support for rafters and roof slope may be provided by a new purlin (dwarf) wall or steel section instead.
- Rafter heads may be fixed to a new roof beam positioned as closely as possible to the apex of the roof, or at a point that provides reasonable headroom.
- Rafters may be fixed to the new floor structure if low-level ties (such as original ceiling joists) are removed. In most cases, the bird's mouthed feet of the rafters will remain fixed to the original wall plate and are not disturbed, particularly on front roof slopes.

## Reasons to remove a purlin

Existing purlins frequently present an obstacle to the loft conversion process in traditional cut and TDA truss roofs, particularly so in small and medium-sized dwellings with a single purlin for each slope:

- The purlin in its existing position is likely to impede dormer construction or may limit headroom on the new stair. On a front (i.e. highway-facing) roof slope, a purlin may foul new window positions.
- The purlin depends on support by strutting or a principal rafter configuration, but these must be removed to increase the useable void as part of the conversion.

## Replacement support for purlins

The purlin's role is to reduce the span of rafters, while the role of the struts or principals is to reduce the span of the purlin. When a purlin and its struts must be removed, some other way must be found to provide support for the roof slope.

### *Purlin wall support for rafters*

Replacement support is generally provided by a low-level structural stud partition (purlin wall). A timber plate is fixed to the new floor joists (or shot-fired to the floor beam) to support timber studs that are, in turn, fixed either to a bearing plate nailed to the underside of the rafters or directly to the existing rafters by bolting (Fig. 9.6). When adopting this approach, a vertical stud is generally provided for each rafter on the slope. Ply may be glued and screwed to the face of the studwork (after insulation has been inserted) to improve rigidity.

The terms *dwarf stud wall* and *knee wall* are also used to describe the new supporting structure. The vertical supporting timber members are sometimes called ashlar studs, although the term should be used with care because ashlar (or ashlaring, sometimes ashlering) is also widely used to describe non-structural eaves infill.

Doors may be provided to allow for eaves storage and access to tanks. Mineral wool pugging and floor decking may be laid directly up to the eaves in order to satisfy fire and sound resistance requirements. In order to simplify the ventilation of the roof structure, thermal insulation material is fixed between the studs of the purlin wall, rather than between lower ends of the rafters where it might interfere with eaves ventilation. Access doors must be insulated if the void is configured as a 'cold' roof space.

**Fig. 9.8**  New purlin wall with remnant of original purlin still in situ.

The critical factor in the positioning of the purlin wall is its relationship with the roof slope it must support. This is determined by the span and spacing of the existing rafters, their sectional dimensions and strength, and roof-loading conditions. Table 9.1 indicates typical maximum spans for a common rafter section size. Note also that the supporting floor structure or supporting beam must be configured to accept a roof loading.

For the practical purpose of providing structural support for the roof slope *before* the purlin is removed, the purlin wall is generally fixed somewhat closer to the eaves than the existing purlin line subject to structural calculations. This has the advantage of providing a greater floor area for the conversion, but there is little practical advantage obtained by a height of less than 600 mm. Note that there are no minimum headroom requirements for rooms.

Support for the roof slope, whether temporary or permanent (e.g. a purlin wall), must always be provided *before* an existing purlin and any of its supporting structure are removed (Fig. 9.8). Even small movements of the existing roof structure caused by alteration are almost impossible to recover once they have occurred.

### Roof beam support for rafters

Typically, a timber ridge board of the sort described earlier is generally intended only to locate rafter pairs in an existing roof. It provides a very limited degree of restraint, and functions in a satisfactory manner only when balancing the rafters of opposing slopes.

In larger dormer conversions, a roof beam spanning from gable to gable must be provided. Typically, this provides a bearing for rafters in the existing roof slope and supports the new dormer's flat roof joists (Fig. 9.9). Depending on the design of the conversion, the beam may be fixed in a number of positions at or near the apex of the roof (Fig. 9.10).

**Fig. 9.9**   Roof beam configuration. Beam infill timbers provide a bearing for rafters (left). Flat roof joists (right) are notched into the web – note slight downstand to allow for shrinkage.

A universal beam or column is commonly used in this application. A flitch beam may also be used at the ridge provided it is adequately restrained against lateral movement.

Timbers fitted into the webbing and bolted or shot-fired through the neutral axis of the beam provide a bearing surface for fixing. Typically, webbing timbers are bolted at 600 mm centres, although it is advisable to ensure that bolt projections do not coincide with fixing points for rafters and joists. Shot-fired fastenings fixed with a powder-actuated tool are increasingly used in this application.

## RAFTERS

### Trimming

Trimming members are required to provide reinforcement for the roof structure when rafters are removed to create an opening (Fig. 9.11). The techniques used to create trimmed openings in roof slopes are broadly similar to those described for trimming floor joists (Chapter 7), but there are variations in framing depending on the type of opening being created. There are three elements common to framing any trimmed roof opening:

- *Trimming rafters* are generally pitched from the wall plate to the ridge, with intermediate purlin or purlin wall support.
- *Trimmed rafters* are the existing common rafters that are cut to accommodate the roof opening.
- *Trimmers* run at right angles to the roof slope and support the cut ends of the trimmed rafters at the head and foot on the opening. Trimmers must be positively fixed to the trimming rafters.

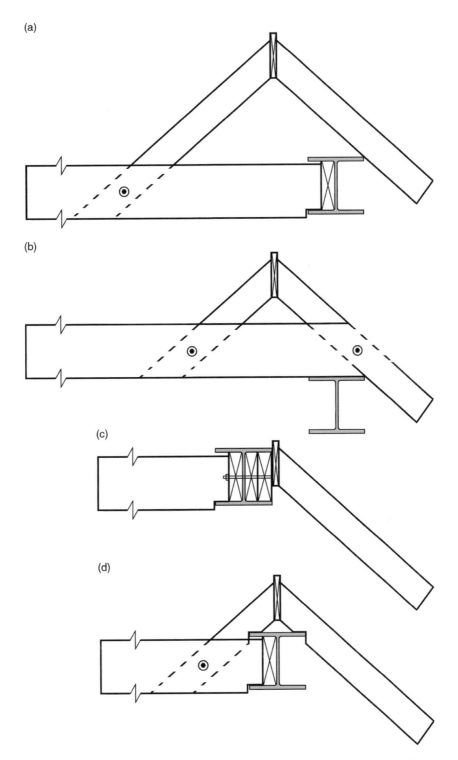

**Fig. 9.10**  Large dormer conversion: roof beam positions. (a) Beam offset, (b) Beam offset –
alternative joist position, (c) Beam at ridge, (d) Beam below ridge with bird's mouth cuts to rafters.

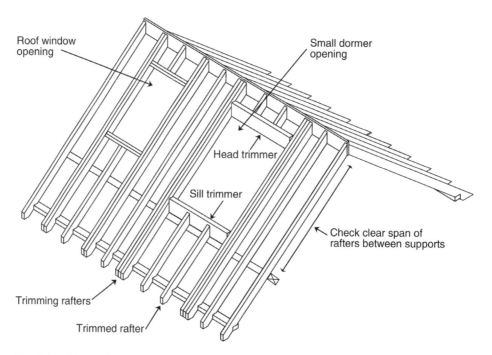

**Fig. 9.11**   Trimmed openings.

## Trimming for roof windows

Structural openings for proprietary roof windows are generally relatively straightforward to create in traditional cut roofs. Note that double and treble trimming rafters may be required for larger windows. Double trimmers may be required to support trimmed rafters at the top and bottom of the opening. Historically, trimmers were fixed to the trimming rafters with pinned tenon joints, with the trimmed rafters dovetailed to the trimmer. In modern practice, timber elements are generally butted and nailed together (Fig. 9.12a). Multiple trimming rafters are bolted together.

## Trimming for dormer projections (general)

The principles for framing a trimmed opening for a dormer projection are similar to those outlined above but there are two key differences. The load of a dormer projection is likely to be considerably greater than that of a roof window occupying a similar area of roof slope, and careful consideration must be given to sizing the trimming rafters. In addition, the relationship between the trimmers and the dormer structure must also be considered in order to simplify construction.

### Small dormer windows

These generally occupy only part of the roof slope. Rafters are trimmed to accommodate the projection. The trimmer at the top of the opening is generally described as a head

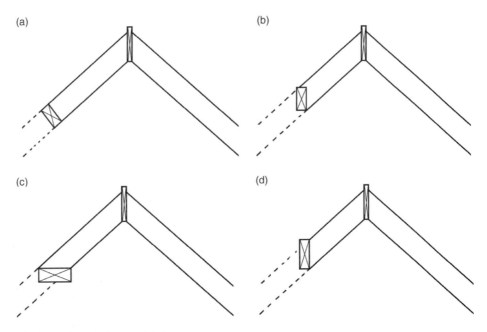

**Fig. 9.12**  Head trimmer detail.

trimmer and this may be oriented in a number of ways depending on the design of the dormer (Figs 9.12b, c and d).

In a small flat roof dormer, the head trimmer may provide direct support for both trimmed rafters and the dormer's flat roof joists. It is therefore configured in a vertical position relative to its axis to facilitate fixing of the dormer roof joists.

The trimmer at the foot of the opening is generally described as the sill trimmer and, like the head trimmer, it may be fixed in a vertical position relative to its axis where a dormer is to be accommodated. It is also generally configured to project above the plane of the rafters to facilitate the fixing of roofing materials and flashings. Historically, trimmed rafters at the head and sill were bird's mouthed over the trimmer; in current practice, they are generally plumb cut and nailed.

The dimensions and numbers of trimming members required for small dormers are determined by several factors. These include the width and length of the opening formed in the roof structure, the new load and the structural support available (e.g. support by purlins or floor structure, see Fig. 9.13). Note that, in larger traditionally constructed dwellings, the roof may be configured with two purlins. These are generally set at one third and two thirds of rafter span, typically on slopes of up to 6 m. In some cases, these are configured to act as trimmers in their own right for original dormer structures.

In many cases, the need to provide a fixing for cheek studs and roof battens to the principal slope means that it may be prudent to provide additional trimming rafters even if there is no prima facie structural engineering case for doing so (Fig. 9.14). Note that in order to accommodate appropriate levels of insulation between studs, it is generally necessary to frame the walls and roofs of small dormer projections in $100 \times 47$ mm studwork (Figs 9.15 and 9.16).

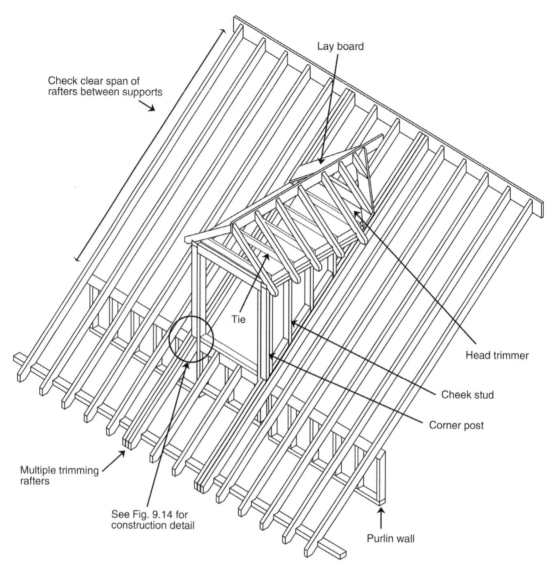

**Fig. 9.13**   Small dormer: construction and trimming.

*Large box dormers*

In cases where a full-width dormer conversion occupies the whole of a pre-existing rear roof slope, there will clearly be no need for a trimmed opening because the entire roof slope is removed. However, many large box dormers are subordinate structures to some degree and retained portions of the existing roof slope – above, below and to the sides of the dormer – must be provided with new support; a conventional trimmed opening alone would usually not be adequate.

- *Lower portion of original roof slope (eaves to base of dormer)*: trimmed rafters may either be fixed to a trimmer which is usually vertically oriented on its axis, or fixed directly to the new box dormer face studwork with bolted or nailed connections.

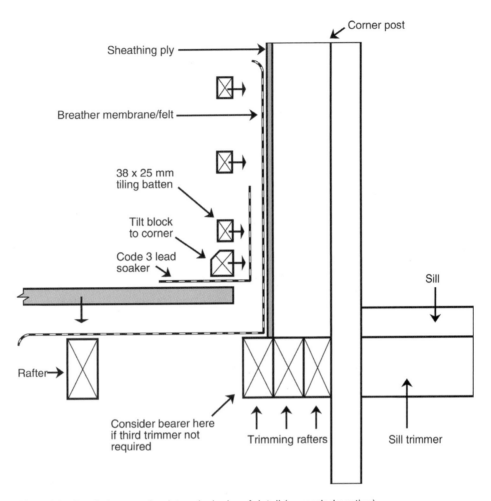

**Fig. 9.14**  Small dormer: cheek to principal roof detail (expanded section).

■ *Upper portion of original roof slope (back of dormer flat roof to existing ridge)*: where a substantial part of the upper slope is to be retained, existing trimmed rafters are fixed directly to new flat roof joists, or to a plate fixed to the flat roof joists. This approach is adopted where one of the primary supports for the new flat roof joists is provided by a load-bearing roof beam (Figs 9.10a, b and d). In many cases, however, only a small amount of the original upper roof slope is left intact and the new flat roof is set at or near the original ridge height.

■ *Side slopes of original roof (eaves to ridge)*: many box dormers are subordinate structures only to the extent that they are formed immediately within the gable walls of the dwelling. The margin of the remaining roof slope flanking the dormer may thus be supported by the gable and in such cases there is no rafter to trim. Where the dormer is narrower, and original rafters retained on one or both sides, these may be reinforced where necessary by the introduction of supplementary trimming rafters. Additional support may also be provided by cheek studwork, where this is supported by a structural floor, or by the provision of a purlin wall.

**Fig. 9.15**    Hipped dormer. Note the provision of a low-level tie.

**Fig. 9.16**    Hipped dormer detail with rigid PIR insulation between studs.

## Sizing and loading of rafters

The complexity of alterations carried out during loft conversions means that it is usually not possible to specify rafters and rafter spacing using span tables. The introduction of trimming loads, repositioning of purlins and the additional loading brought about by the introduction of insulating material and plasterboard finishes to the inside of slopes means that justification for rafter spacing must be made by calculation.

Where new rafters are introduced to reinforce a roof slope, they must generally be of the same section size as the existing rafters in order that they may be accommodated within the thickness of construction. It should be noted that rafter spacing is often not equal in older buildings. Table 9.1 is provided for illustration purposes only.

## HIP-TO-GABLE CONVERSION

Hip-to-gable conversions are carried out in order to increase the volume and useable full-headroom floor area in a loft conversion. An existing section of hipped roof is removed (Fig. 9.17) and the remaining part of the original roof extended to meet a new gable wall. The new gable may be constructed using masonry or timber studwork (Fig. 9.18). Note that a correctly configured studwork gable wall is capable of providing support for one or more steel roof beams.

On the roof slopes that are retained (generally only the front slope but occasionally the rear roof slope as well), the jack rafters and existing roof covering are sometimes left in situ. However, in smaller dwellings, the removal of the hip may mean very little of the original roof remains. Slates or tiles recovered from the redundant hip may provide a source of well-matched roofing material for the newly extended slope or slopes.

**Fig. 9.17**   Hip-to-gable conversion: positioning new roof beam.
Courtesy South London Lofts Ltd.

**Fig. 9.18** Hip-to-gable conversion: new gable wall with plain tiling.
Courtesy South London Lofts Ltd.

As with any structural alteration, careful attention should be given to the provision of both temporary and permanent supports while work is being carried out.

## NOTCHES AND HOLES

It may be necessary to provide structural engineering calculations to justify cutting holes or notches into elements of roof structure, for example, to permit continuous ventilation around roof windows (Fig. 9.19). Note that only floor and flat roof joists may be drilled or notched without formal calculation, as long as the guidance reproduced in Chapter 7 is followed (Fig. 7.8).

## LATERAL SUPPORT FOR GABLES

In order to resist forces acting horizontally (such as wind loading), guidance in Approved Document A *Structure* is that gable walls should be strapped to the roof structure. When a loft is converted, the extent of compliance with this guidance depends on the nature of the alterations to the roof. In most traditionally constructed cut roofs, restraint fixings were not provided and, in some cases, it may be prudent to fit them as part of the conversion even if it is not a requirement.

- *Existing pitched roof slope unaltered.* There may not be a requirement to retro-fix lateral restraint strapping if a roof slope is substantially unaltered by building work, for example in the case of a front roof slope to a gable-ended dwelling.

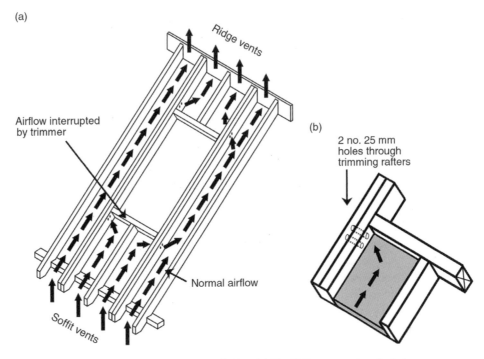

**Fig. 9.19**   Insulated cold roof slopes: ventilation. (a) Ventilation around roof window, (b) Drilling detail below trimmer.

- *Hip-to-gable conversion.* Where an entirely new gable is built to replace a roof hip, either in masonry or studwork, it is necessary to provide lateral tension straps between the new wall and the new roof structure. This applies equally to the construction of a flank gable wall (see also Chapter 8).
- *Provision of lateral support for gable walls.* A method of providing lateral support for gable walls is described in Approved Document A. Tension straps (generally galvanised mild steel $30 \times 5$ mm) are fixed between the roof slope and the gable wall. A strap is provided at or near the highest point of the roof, and additional straps fixed at no more than 2 m centres down the roof slopes.

## REPLACEMENT ROOF COVERINGS

In cases where it is proposed to replace the roofing material on an existing roof with one of significantly greater mass, the structural integrity of both the roof and its supporting structure must be checked. Approved Document A *Structure* indicates that 'significant' in this context means an increase in roof loading of 15% or more.

For example, replacing slates (unit load approximately $0.30 \, kN/m^2$) with concrete tiles (unit load approximately $0.51 \, kN/m^2$) would result in a 'significant' change in load.

The guidance in Approved Document A is that the roof structure and supporting structure should be checked to ensure that on completion, the building is not less compliant with Requirement A1 than the original building. Note that any strengthening work or replacement of roof members that might be required would be classified as a material alteration.

Also, where a replacement roofing material has a *lower* mass than the original, the roof structure and its anchoring should be checked to ensure that an adequate factor of safety is maintained against the risk of wind uplift.

## FLAT ROOF: BASIC STRUCTURE

Flat roofs – either 'warm' deck or 'cold' deck – are an element common to all box dormer conversions. Typically, flat roof joists are fixed to a roof beam shared with the existing pitched roof at or near the ridge (Fig. 9.10) and are strapped to the header at the dormer face (Fig. 9.20). The slope of the flat roof is generally provided by tapered timber firring pieces fixed to the top of the roof joists and is laid to fall to the rear of the building (i.e. towards the dormer face) where guttering must be provided.

Strutting should be provided between roof joists – see Chapter 7, *Strutting*. Flat roof joists may be selected from *Eurocode 5 span tables for solid timber members in*

**Fig. 9.20** Box dormer. Sheathing and roof deck in 15 mm OSB. Flat roof joists are strapped to the header; inverted 'long-leg' hangers have been used in this example.

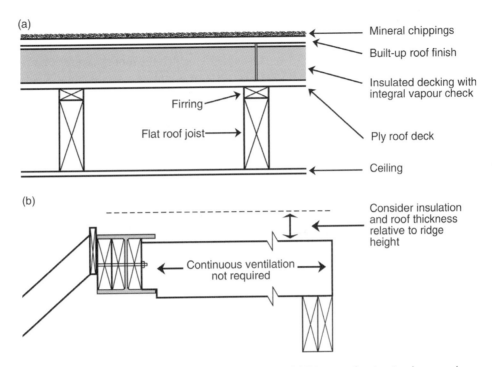

Fig. 9.21 Flat roof: warm deck. (a) Warm roof section, (b) Warm roof – structural approach.

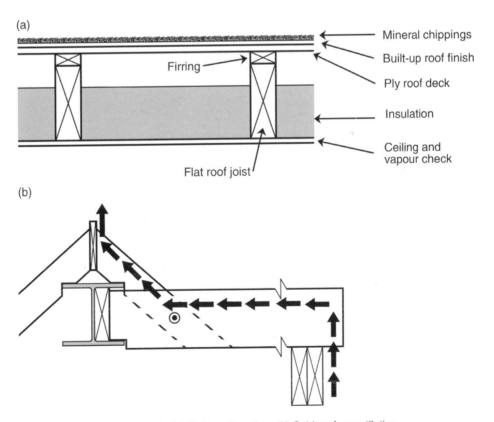

Fig. 9.22 Flat roof: cold deck. (a) Cold roof section, (b) Cold roof – ventilation.

**Table 9.2**  Flat roof joists (C16) – permissible clear spans.

| Joist size (mm) | Joist spacing (mm) | Clear span (m) |
|---|---|---|
| 47 × 147 | 400 | 2.96 |
| 47 × 170 | 400 | 3.53 |
| 47 × 195 | 400 | 4.14 |
| 47 × 220 | 400 | 4.75 |
| 75 × 147 | 400 | 3.56 |
| 75 × 170 | 400 | 4.21 |
| 75 × 220 | 400 | 5.50 |

Maintenance/repair access only.
Imposed snow load 0.75 kN/m$^2$.
Dead load 0.75–1.00 kN/m$^2$ (excluding self-weight of joist).
For general guidance only. Source: Approved Document A (1992).

**Table 9.3**  Flat roof joists (C24) – permissible clear spans.

| Joist size (mm) | Joist spacing (mm) | Clear span (m) |
|---|---|---|
| 47 × 147 | 400 | 3.11 |
| 47 × 170 | 400 | 3.69 |
| 47 × 195 | 400 | 4.33 |
| 47 × 220 | 400 | 4.94 |
| 75 × 147 | 400 | 3.73 |
| 75 × 170 | 400 | 4.40 |
| 75 × 220 | 400 | 5.70 |

Maintenance/repair access only.
Imposed snow load 0.75 kN/m$^2$.
Dead load 0.75–1.00 kN/m$^2$ (excluding self-weight of joist).
For general guidance only. Source: Approved Document A (1992).

*floors, ceilings and roofs for dwellings* (TRADA Technology Ltd). An indication of permissible clear spans is given in Tables 9.2 and 9.3. These are based on the span tables included in the 1992 version of Approved Document A and are provided for illustration only.

## Flat roof – warm deck (unventilated)

In a flat warm roof, the insulating material is carried above the roof joists (Fig. 9.21a). Because the voids between the ceiling finish and the underside of the insulation are normally at more or less the same temperature as the room below them, the risk of moisture-laden air condensing within the structure is reduced and ventilation of the void is therefore not required.

Because the insulation is fixed on top of the roof joists, rather than between them, a warm flat roof is thicker than a cold one. This is sometimes problematic in conversions where headroom is limited and where planning restrictions place limits on roof height (Fig. 9.21b). Note, however, that this system of roof construction offers a number of practical benefits:

■ Because there is no need to ventilate the void, a warm roof can be used in applications such as dormer construction where cross ventilation is not always possible.

- Thermal bridging is eliminated and the roof will provide more effective insulation.
- Construction is quicker and less labour intensive.
- Recessed lighting (downlights) can be fixed without disturbing the insulation; cables need not be derated because they do not run through insulation.

## Flat roof – cold deck (ventilated)

Insulation in a cold flat roof is accommodated in the void between the roof joists and in many cases, below the joists as well. This results in a roof that has a relatively shallow section (compared to the warm deck) which is advantageous in a conversion where ceiling height is restricted (Fig. 9.22a).

A cold flat roof, sometimes called a cold deck, requires cross ventilation to reduce the risk of water vapour from within the building condensing within or above the insulating material (Fig. 9.22b). A vapour check is provided between the ceiling and insulation to further minimise this risk. The following factors should be considered:

- Air for ventilation must be allowed to flow freely through the 'cold' part of the flat roof. Eaves ventilation is generally provided at the dormer face by oversailing the flat roof joists to create space for soffit vents. Where the existing ridge structure and part of the original roof slope are retained, it is possible to achieve continuity of airflow to a new or existing ridge or high-level vent. In cases where the joists at the ridge end of the flat roof must be supported by a beam that is fixed at ridge level (Fig. 9.10c), providing front-to-back cross-flow ventilation may not be possible. Warm deck construction must be considered in these cases, but as noted above, the finished height of such a roof must be considered relative to planning restrictions.
- Because insulation is carried between the roof joists, rather than above them, a degree of thermal bridging will occur. Supplementary ceiling-level insulation may therefore be required.
- The use of recessed lighting creates difficulties where a cold roof is to be used. To limit localised overheating caused by lamps, it is necessary to remove surrounding insulation; cables may need to be derated where they pass through insulation. In addition, any recessed fittings will breach the ceiling vapour check and allow moisture-laden air from the rooms below into the void.

## Flat roof – hybrid warm roof (unventilated)

In some cases, it is possible to adopt a hybrid approach, with insulation above and between roof joists. Because the roof is not ventilated and because some of the insulation is accommodated within the joist voids, it results in a thinner overall section, but it is not universally accepted by local authorities (see Fig. 10.8b).

## Roof ventilation

Ensuring that a cold roof structure (flat or pitched) is resistant to damage caused by condensation is a requirement of the Building Regulations and, as noted above, it is generally necessary to provide ventilation. In order to reduce the risk of moisture-laden air from inside a dwelling condensing within the structure (interstitial condensation), a dual strategy is adopted to protect the roofs of dwellings:

- A vapour check is provided between internal plasterboard and the warm side of the insulation to limit the passage of water vapour from inside the building to the roof structure. This may take the form of vapour-impermeable sheet material, such as polythene, or wallboards with an integral vapour check.
- Any moisture-laden air that does enter the roof structure must be allowed to escape. This may be achieved by providing a 50 mm ventilation void between the cold side of the insulating material and the underside of the existing vapour-impermeable roofing felt or roof deck. In pitched roofs where the ceiling follows the roof pitch, airflow through this void is provided by openings equivalent to 25 mm at the eaves (continuous) and 5 mm at the ridge (continuous).

Breather membrane may be provided instead of traditional underlay. Breather membranes allow water vapour to escape but prevent the ingress of liquid water and have the advantage of being easier to handle than traditional underlays.

## Ventilation for pitched cold roofs

A 'cold roof' in the context of a loft conversion is one where the insulating material is carried beneath or between the rafters (the expression is also used to describe an un-converted roof void where insulation is provided at ceiling level). A characteristic of most loft conversions is that the ceiling follows the pitch of the roof, at least at some point, and this generally applies to a large portion of the slope nearest the highway. To reduce the risk of interstitial condensation occurring in situations such as this, a 50 mm gap must be provided between the upper (cold side) of any insulation and the underlay. In tandem with this, ventilation must be provided at the eaves (equivalent to a 25 mm continuous strip) and at the ridge (equivalent to a 5 mm continuous strip).

Providing ventilation at the apex of the roof is generally a relatively straightforward matter of fitting proprietary ridge vents; airflow at the eaves may be achieved through the use of soffit vents. However, in the case of dwellings with flush eaves (Fig. 9.3), providing airflow is sometimes more troublesome. Where soffit or fascia vents cannot be accommodated, proprietary tile or slate vents may be provided. In such cases, it is best that insulation inside the conversion is carried between the studs of the purlin wall (thus creating a continuous cross-void at eaves level) rather than following the roof slope down to the rafter feet. If insulation were to be carried down to the eaves, it would be necessary to provide a vent for each rafter void, which is both costly and unsightly.

## Ventilation for pitched warm roofs

A warm roof is one where the insulating material is carried above the rafters. Because the air within any structural voids is similar in temperature to the room beneath, the risk of condensation occurring is reduced. It is not necessary to ventilate warm roof voids.

Pitched warm roofs are a relatively recent innovation. Note that it is seldom practical to provide warm roof insulation to existing pitched roofs in terraced or semi-detached dwellings because it increases the outward projection of the roof envelope relative to adjoining dwellings. An out-of-plane alteration to the roof of this sort is likely to fall foul of permitted development rules.

### Ventilation for flat roofs – cold deck

As noted above, the ventilation provisions for cold deck flat roofs are similar to those for pitched roofs where the ceiling finish follows the roof pitch (above). A 50 mm gap is provided between the cold side of the insulation and the roof deck, and airflow is provided by the equivalent of continuous 25 mm ventilation at the dormer face and 5 mm at the ridge. Where this cannot be achieved, an alternative is to provide a warm deck roof.

### Ventilation for flat roofs – warm deck

As noted earlier, there is relatively little risk of condensation forming within the roof void and it is therefore not necessary to provide ventilation.

## Continuity of airflow around roof windows

In cold-pitched roofs, the provision of roof windows will block the ventilation path, and steps must be taken to ensure that adequate airflow is maintained. Where a roof window is to be fixed in an original roof slope, one way of achieving continuous ventilation is to drill 25 mm diameter holes through the trimming rafters immediately above and below the trimmed opening (Fig. 9.19).

In wider openings, it is necessary also to drill through the intervening trimmed rafters. To be effective, these holes must be drilled *above* the level of insulating material between the rafters. Note that it would generally be necessary to justify any such proposal by calculation. Alternatively, external vents may be provided immediately above and below the obstruction.

## Ventilation – possible exemptions from the requirement

Approved Document C states that:

> For the purposes of health and safety it may not always be necessary to provide ventilation to small roofs such as those over porches and bay windows.

It is not clear to what extent this guidance would apply to a section of pitched or flat roof on a small dormer projection. It is generally prudent to work on the assumption that *all* cold roofs require ventilation. Any proposal not to ventilate should be the subject of discussion with, or made as part of a full-plans submission to, the building control service chosen.

## Approved Document guidance

Guidance on protecting roofs from condensation was originally included in Approved Document F2. In 2004, however, the guidance was moved to Approved Document C2. Advice on methods of ventilating roof voids has been removed and, instead, Approved Document C indicates that the requirement to resist damage from interstitial condensation will be met by designing a roof in accordance with clause 8.4 of BS 5250:2002 and BS EN ISO 13788:2002, with further guidance provided in BRE Report BR 262.

## ATTIC TRUSSES

The trussed rafter roof (Fig. 9.23a) was first demonstrated in Britain in 1963 and was rapidly adopted for domestic roof construction during the 1970s. It remains the dominant method of domestic roof construction in the UK for new build.

Roofs made with conventional trussed rafters (these are sometimes described as 'W' or 'Fink' trusses) are difficult to modify and conversion is seldom straightforward. However, the trend towards higher housing densities has led to a radical reappraisal of the way roof voids are used. Today, more than a third of new dwellings incorporate attic or RiR trusses (Fig. 9.23b). These yield a useable roof void and are designed for domestic floor loading (Table 9.4). Houses constructed with RiR trusses may be adapted for immediate habitable use, or fitted out at a later date.

Larger attic trusses – those over 4 m in height – may be constructed in two parts to facilitate transportation and handling. Fig. 9.24 illustrates a hybrid RiR arrangement, with a traditional 'cut' lower section and prefabricated upper section in a new-build project.

It is possible to use attic trusses to completely replace the roof structure of an existing building and this approach is gaining popularity. There are two principal approaches. One is to remove the existing roof and ceiling structure in a single operation, although clearly removing ceilings is a highly disruptive procedure. The other method is to replace existing trusses with new attic trusses one at a time, tying the existing bottom chords (which carry the ceiling) to the new trusses as work progresses.

The main drawbacks with a total roof replacement are likely to be high crane handling costs, and the need to provide temporary storage for the new trusses when they are delivered.

**Table 9.4**   Room-in-roof trusses – minimum practical dimensions.

| Roof pitch | Minimum span | Attic room width between roof support stud walls |
|---|---|---|
| 35° | 9 m | 4.5 m |
| 40° | 7 m | 4.0 m |
| 45° | 6 m* | 3.5 m |
| 50° | 6 m* | 4.0 m |

*Spans of 6 m or more may require intermediate support from below.

(a)

(b)

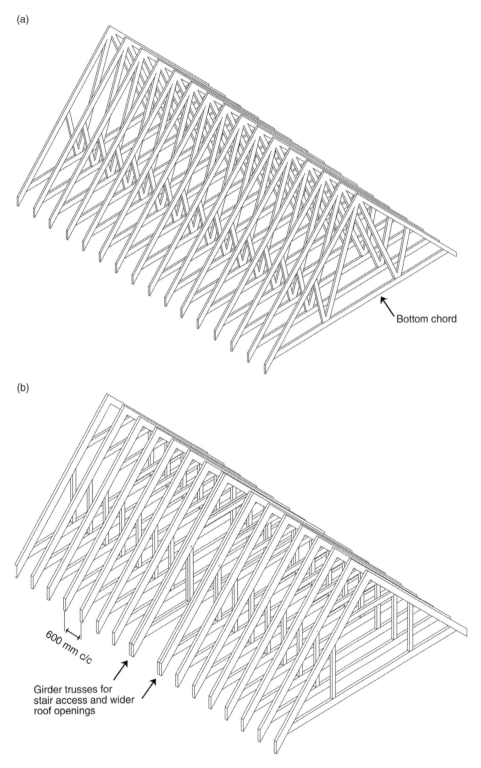

Bottom chord

600 mm c/c

Girder trusses for
stair access and wider
roof openings

**Fig. 9.23**   Trussed rafter roofs (stability bracing and infill omitted). (a) Conventional webbed truss, (b) Attic truss.

**Fig. 9.24**   Hybrid new-build attic with hand-cut rafters (lower slope) and prefabricated 'top hat' trusses.

Note also that the arrangement of sole plates and wall plates in the roof to be converted must be perfectly regular in order to accept the new trusses. Dimensional regularity in older masonry-constructed dwellings should not be assumed and trusses are generally intolerant of even minor modifications.

# 10 Energy performance

The legal requirement to conserve fuel and power is set out in Part L of Schedule 1 to the Building Regulations 2010. Guidance on complying with Part L and other related Building Regulations requirements is provided in Approved Document L *Conservation of fuel and power*. Additional guidance is provided in the *Domestic Building Services Compliance Guide*, the *Domestic Ventilation Compliance Guide* and *Accredited Construction Details*.

Energy performance requirements continue to represent a moving target: a revised version of Part L and new Approved Document guidance came into effect on 1 October 2010; further revisions – with potentially tougher targets – are planned for 2013 and 2016. Note that the 2010 Approved Document is written with reference to the Building Regulations 2000 (as amended) which has now been revoked, rather than the new consolidated version (the Building Regulations 2010), although both the new Approved Document and new Building Regulations were published on the same day.

Approved Document L (2010) is arranged in four parts, with work in existing dwellings covered in document L1B (ADL1B). This document concerns four types of work in existing dwellings:

- The construction of an extension
- A material change of use, or a change to the building's energy status, including such work as loft conversions and garage conversions
- The provision or extension of a controlled service or controlled fitting
- The replacement or renovation of a thermal element

Note that a loft conversion in a single family dwellinghouse would not normally constitute a 'material change of use' for the purposes of the Building Regulations.

## METHODS OF COMPLIANCE

ADL1B describes three approaches to demonstrating compliance. These are described in the following text.

### The reference method (elemental approach)

This is the primary route to achieving compliance when a dwelling is extended and following the guidance is relatively straightforward. With this method, a minimum thermal transmittance value – U-value – is assigned to each of the following elements within the extension. These are referred to as 'fabric standards' and they apply to:

- Roofs
- Walls
- Floors
- Windows and rooflights
- Doors

The reference method also sets out guidance on heating, lighting and the extent of external glazed areas and doors. Because this method considers individual building elements, rather than whole-building emissions, it is sometimes described as the *elemental approach* or *elemental method*, although it is not defined as such in the current Approved Document.

## Area-weighted U-value method (optional approach)

This is a variant of the 'reference method' described above. It offers a more flexible approach to compliance because it allows modest trade-offs between elements. Under the area-weighted method, a lesser standard may be applied to some elements in the extended dwelling (e.g. a reduction in roof insulation), provided that proportionate compensation is made elsewhere in the extension (e.g. by providing better wall insulation).

To prevent a very cold element (which would increase the risk of localised condensation) being traded off against additional insulation elsewhere, limiting U-values apply. Although no longer specifically referenced in ADL1B, these are the same as the threshold values in Table 10.1, column (a).

**Table 10.1**   Upgrading retained thermal elements.

| Element | (a) Threshold U-value (W/m² K) | (b) Improved U-value (W/m² K) |
|---|---|---|
| Wall – cavity insulation* | 0.70 | 0.55 |
| Wall – external or internal insulation† | 0.70 | 0.30 |
| Pitched roof – insulation at ceiling level | 0.35 | 0.16 |
| Pitched roof – insulation between rafters‡ | 0.35 | 0.18 |
| Flat roof or roof with integral insulation§ | 0.35 | 0.18 |
| Floor‖ | 0.70 | 0.25 |

**Notes**
- 'Roof' includes roof parts of dormer windows.
- 'Wall' includes wall parts (cheeks) of dormer windows.

*Applies only in the case of a wall suitable for the installation of cavity insulation. If this is not the case, it should be treated as 'wall – external or internal insulation'.

†A lesser provision may be appropriate where meeting such a standard would result in a reduction of more than 5% in the internal floor area of the room bounded by the wall.

‡A lesser provision may be appropriate where meeting such a standard would create limitations on headroom. In such cases, the depth of the insulation plus any required air gap should be at least to the depth of the rafters, and the thermal performance of the chosen insulant should be such as to achieve the best practicable U-value.

§A lesser standard may be appropriate if there are particular problems associated with the load-bearing capacity of the frame or the upstand height.

‖In a typical loft conversion, a floor would not require thermal insulation.

## Whole dwelling calculation method (optional approach)

This method is based on whole-building $CO_2$ emissions with calculations carried out using the government's Standard Assessment Procedure set out in SAP 2009. The emission rate for the existing dwelling, plus its proposed extension, must be *no greater* than that of the existing dwelling with a notional extension that conforms to the fabric standards in the reference method (described above).

As with the area-weighted approach, limiting U-values (Table 10.1, column (a)) apply to individual elements to minimise condensation risks (detail is contained in Approved Document C). The 'whole dwelling' method provides additional flexibility because it allows trade-offs to be made throughout the dwelling, but it is a more complex approach than the reference and area-weighted approaches described above.

## WALLS AND ROOFS – PERFORMANCE REQUIREMENTS

A wall, roof or floor that forms part of the thermal envelope of a building is described as a *thermal element* in the Building Regulations 2010 and ADL1B.

Note that a roof with a pitch of more than 70° (such as that forming the lower slope of a mansard roof) should be insulated as if it were a wall.

The standard of thermal performance required for walls and roofs is expressed as a U-value and it is dependent on whether the element in question is 'new' or 'existing'; in the case of 'existing' elements, the U-value required depends on the type of work being carried out. ADL1B classifies thermal elements in the following way:

- New thermal elements
- Existing thermal elements – retained
- Existing thermal elements – replacement
- Existing thermal elements – renovated (renovation of a thermal element)

In most cases, the construction of a new box dormer loft conversion will comprise both 'new' thermal elements (such as the walls and roof of the new dormer structure) and 'existing' elements (such as an existing front roof slope and gable walls).

Although ADL1B treats 'existing' thermal elements in three different ways, the guidance indicates that such elements in a new loft conversion can be treated as 'retained' thermal elements, that is, elements that will become part of the thermal envelope of the building where previously they were not.

When a new loft conversion is built, therefore, a mixture of two sets of U-value standards will normally apply: the standard for 'new' thermal elements and the standard for 'retained' thermal elements. Fig. 10.1 illustrates a typical configuration with values drawn from the two different standards.

Note that the floor of a loft conversion forming an additional storey in a single-occupancy dwellinghouse is unlikely to require any thermal insulation for the purposes of part L because it is not normally part of the building's thermal envelope. However, mineral wool is generally integrated within the new floor structure to meet the requirements of parts B (fire resistance) and E (sound resistance) of the Building Regulations.

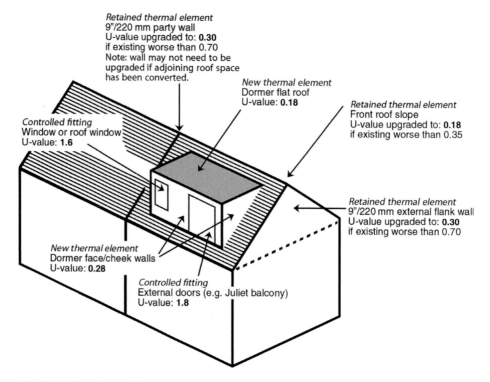

**Fig. 10.1**   Box dormer loft conversion: elemental standards.

## U-values for new thermal elements

In a typical dormer conversion, the fabric standards for 'new thermal elements' would apply to the following:

- New dormer walls, including cheeks
- New dormer roof
- New gable or flank wall (e.g. as part of a hip-to-gable conversion)
- New 'knee', 'dwarf' or 'purlin' stud wall built at the foot of an existing roof slope
- New sections of pitched roof

## U-values for retained thermal elements

In the case of a dormer loft conversion, 'retained' thermal elements are those elements bounding the original roof void that become part of the conversion's thermal envelope. These might include some or all of the following:

- Existing sections of pitched roof
- Existing gable walls
- Existing party walls

Existing thermal elements that already meet the 'threshold' U-values in column (a) of Table 10.1 need not be modified. In practice, however, the extremely poor performance of

**Table 10.2**  Existing thermal elements (un-insulated) – some common values.

| Element | Unmodified U-value (W/m² K) |
| --- | --- |
| Cavity wall (concrete block inner leaf) | 1.6 |
| Solid brick masonry wall (220 mm) | 2.1 |
| Pitched roof – cold (tile/slate) | 4.9 |
| Floor (e.g. over unheated garage) | 1.9 |

typical as-existing elements encountered in loft conversions (such as un-insulated pitched roof slopes and gable walls) will trigger the need to meet the 'improved' U-values in column (b) in most cases.

Table 10.2 indicates possible values for un-insulated thermal elements that might be encountered in the roof void *before* conversion. It should be noted that a considerable degree of variation in thermal performance is possible, particularly with cavity walls. Additional guidance on the thermal performance of as-built elements is set out in *The Government's Standard Assessment Procedure for Energy Rating of Dwellings* (SAP 2009), Appendix S.

ADL1B recognises the difficulties inherent in improving the performance of existing fabric: upgrading should therefore be carried out where it is technically, functionally and economically feasible. A test of economic feasibility is provided in the Approved Document guidance, that is, a simple payback of 15 years (see also *simple payback*, Glossary).

The guidance suggests a technical argument for lesser provision would apply if the existing structure were not capable of supporting the weight of additional insulation. ADL1B also indicates that a lesser provision might apply where the thickness of the insulation would reduce useable floor area by more than 5%. The lesser standard should generally not be worse than 0.70 W/m² K.

## Standards for replacement thermal elements in an existing dwelling

In many cases, these standards (the same as those for new thermal elements, Table 10.3) would not apply in the context of a new loft conversion, unless an existing element (such as a gable wall or roof slope) were to be entirely removed and replaced with an identical new element in the same position. These standards would also apply if, say, an existing dormer structure were to be replaced on a like-for-like basis.

## Standards for renovation of thermal elements

These standards apply when certain types of work are carried out in an *existing* loft conversion or room-in-roof configuration. They are triggered when more than 50% of the surface area of the individual element (or 25% of the total building envelope) is renovated. The values set out in Table 10.1 column (b) would apply in these cases, where feasible.

**Table 10.3**    Fabric standards: U-values for new thermal elements.

| Element | Standard (W/m² K) |
|---|---|
| Wall* | 0.28 (area-weighted average values) |
| Pitched roof – insulation at ceiling level† | 0.16 |
| Pitched roof – insulation at rafter level‡ | 0.18 |
| Flat roof or roof with integral insulation | 0.18 |
| Floors§ | 0.22 |

**Notes**
• 'Roof' includes roof parts of dormer windows.
• 'Wall' includes wall parts (cheeks) of dormer windows.
*A lesser provision may be appropriate where meeting such a standard would result in a reduction of more than 5% in the internal floor area of the room bounded by the wall.
†This refers to insulation laid over ceiling joists (i.e. at floor level) in a roof space and would not apply in a typical loft conversion.
‡This refers to the sloping ceiling formed by adaptation of a roof slope and is relevant in most conversions.
§In a typical loft conversion, a floor would not require thermal insulation.

## ENERGY CONSERVATION – PRACTICAL APPROACHES

The following notes and drawings are intended to provide an outline of some of the practical measures that can be taken to limit heat loss through walls, roofs and windows in loft conversions. Both new and retained thermal elements are considered.

It is emphasised that this section is intended to provide general guidance only. In individual cases, it will be necessary to carry out a thorough assessment of the existing building fabric; declared performance standards and installation guidance provided by insulation manufacturers should also be taken into account. Note also that major manufacturers are able to provide technical specifications for specific constructions.

### Insulation materials

Headroom and floor space are generally at a premium when lofts are converted. So to keep the thickness of insulating materials to a minimum, it is practice to use high-performance foil-faced polyisocyanurate (PIR), polyurethane (PUR) or phenolic foam insulation boards in the majority of applications. These rigid, low-mass materials offer a high level of thermal resistance relative to their thickness: a 12 mm thickness of PIR insulation on its own, for example, has better thermal resistance than a 9" brick wall.

A number of manufacturers produce insulated plasterboard or thermal laminate. This is a composite material comprising a layer of rigid insulation (e.g. PIR or PUR) bonded to a plasterboard facing. The plasterboard is typically either 12.5 or 9.5 mm in thickness, with insulation thicknesses typically ranging from 25 to 65 mm.

In most cases, a continuous vapour control layer (VCL) must be provided across and between elements. This may take the form of polythene sheeting; equally, it may be integral with proprietary wallboards or rigid insulation materials with continuity maintained by taping or the application of a mastic bead. Manufacturers' guidance should be followed.

Note that low-emissivity cavities in insulated constructions (e.g. Fig. 10.2c) make a contribution to limiting heat loss from the building envelope.

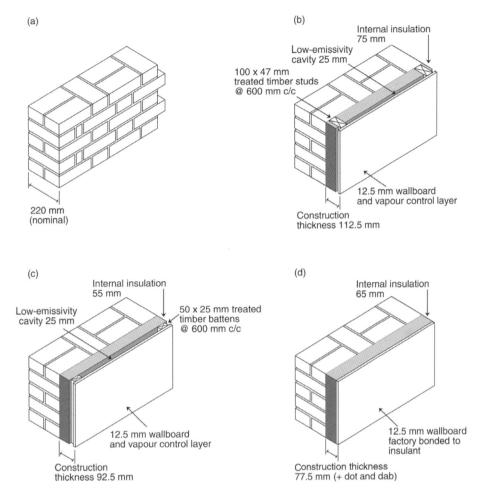

**Fig. 10.2** Upgrading solid brick masonry walls: internal insulation with rigid PIR foam boards. U-values and construction details are for guidance only – insulant manufacturer's data should be consulted in all cases. (a) Solid 9" brick wall (unmodified) assumed U-value: 2.1 W/m²K, (b) Insulation between studs potential U-value: 0.28 W/m²K, (c) Battens over insulation potential U-value: 0.28 W/m²K, (d) Thermal laminate potential U-value: 0.28 W/m²K.

Non-rigid insulating materials – including mineral wools and multi-foil products – may also be used in loft conversions, although the thickness of mineral wool needed to achieve the required U-values means its application is limited to some extent. Flexible multi-foil insulation, however, is becoming increasingly popular in certain applications, including the insulation of cold-pitched roofs.

## Fixing internal insulation

Generally, the only practical way of doing this is to provide insulation to the inside of the wall. It is also possible to fix insulation *outside* the building envelope, but this is usually not feasible unless the whole height of a wall is similarly treated.

To save space, insulation for walls (and roofs) in loft conversions is often slotted between repeating structural components, such as timber wall studs, roof/ceiling joists and rafters. A typical terraced house conversion will have in excess of 5 m³ structural voids that can be insulated in this way.

The following three approaches are appropriate for brickwork, blockwork and cavity walls where insulation is to be provided internally:

***Insulation between studs***. Treated timber studs are fixed directly to the wall, typically at 400 or 600 mm centres to coordinate with plasterboard sheets. Insulation is then inserted between the studs (e.g. Figs 10.2b and 10.3b). A continuous VCL and plasterboard are then fixed.

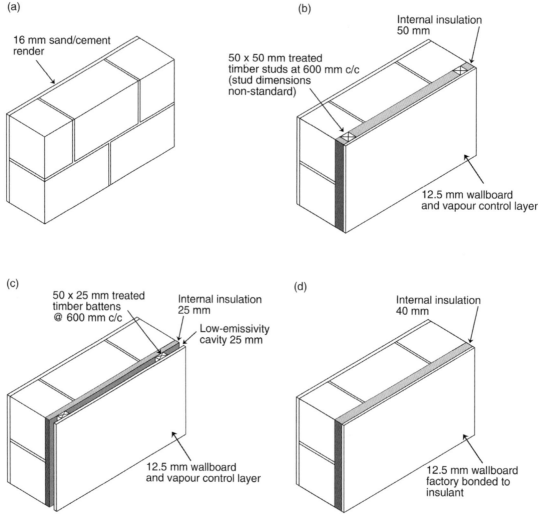

**Fig. 10.3**   Upgrading 215 mm aerated concrete blockwork walls (block conductivity: 0.11 W/m K) – internal insulation with rigid PIR foam boards. U-values and construction details are for guidance only – insulant manufacturer's data should be consulted in all cases. (a) 215 mm blockwork (unmodified) typical U-value: 0.46 W/m² K, (b) Insulation between studs potential U-value: 0.28 W/m² K, (c) Battens over insulation potential U-value: 0.28 W/m² K, (d) Thermal laminate potential U-value: 0.28 W/m² K.

The advantages with this approach are that (a) the studwork provides a robust structure for swift and secure fixing of plasterboard during construction and subsequently applied minor loads, such as bookshelves, and (b) it provides a relatively shallow thickness of construction, thereby saving floor space. The disadvantage is that the insulation is not continuous, with studwork creating thermal bridging of around 15%.

***Continuous insulation with battens over***. Insulation is held in place by timber battens (50 × 25 mm) that are fixed through the insulating board through to the wall. Plasterboard is then fixed directly to the battens to leave a 25 mm cavity (Figs 10.2c and 10.3c).

The advantages with this approach are that (a) it eliminates thermal bridging and (b) it reduces cutting wastage of the insulant because studwork is absent. The disadvantages with this approach are that (a) it can be troublesome to fix the timber battens through insulant into brickwork with any degree of certainty and (b) the 25 mm cavity creates a marginally thicker construction than the stud method described above.

***Thermal laminate***. Composite boards comprising insulant bonded to wallboard are fixed to brickwork with a combination of dot and dab plaster and mechanical fixings (Figs 10.2d and 10.3d).

The advantages with this approach are (a) it provides insulation and wallboard in a single fix and (b) it provides a relatively shallow depth of construction. The disadvantages are that boards are relatively cumbersome and therefore awkward to handle in a confined space. Thermal laminate is best-suited to large uninterrupted expanses of wall, rather than the complex and awkwardly shaped corners and returns that might be found in a loft.

## Airtightness

Although airtightness is not measured in loft conversions, Approved Document L1B indicates that reasonable provision should be made to reduce unwanted air leakage through the building fabric.

In addition to the discomfort it causes for occupants, uncontrolled air leakage is undesirable for two main reasons. Firstly, it increases energy consumption – gaps and cracks allow heated air to escape and cold air from outside to enter the building. Secondly, it allows warm, damp air escaping through cracks to condense within the building fabric (interstitial condensation). This can reduce the effectiveness of insulating materials and damage the building fabric.

In general, no special materials are needed to ensure airtightness: the primary air barrier in most cases is simply plaster and plasterboard used to line walls and ceilings; in the case of loft conversions, airtightness is enhanced by external sheathing to stud walls (see Fig. 8.5). The continuity of the internal (i.e. plasterboard) air barrier at awkward junctions is improved by using tapes or flexible sealants. Attention should be paid to sealing the following potential sources of air leaks:

- Dormer cheeks (complex angles and returns require additional attention)
- Floor/wall junctions (e.g. new dormer/existing rear wall)
- Socket outlets and switch plates (particularly those in external stud walls)
- Service penetrations (especially those concealed behind bath panels and shower trays)

- Window frames
- Window trickle ventilators (these should be airtight when closed)
- Recessed light fittings (consider air-sealed fittings)
- Extract fan outlet ducts

It is emphasised that airtightness measures are intended to minimise *uncontrolled* air leakage. Controlled air flows are, of course, essential to meet the requirements of parts F and J of the Building Regulations. Additional guidance on airtightness is provided in *Accredited Construction Details*, published by the Department for Communities and Local Government.

## Thermal bridging

Thermal bridges (cold bridges) are points or areas in the building envelope where heat losses occur because the continuity of insulation is interrupted.

Bridging may occur over extensive areas in cases where insulation is slotted between structural components, such as wall studs, roof/ceiling joists and rafters. Evidence of bridging – discoloured lines or patches that correspond with timber positions where no insulation is present – may occur many months or even years after final finishes have been applied.

The bridging effect is exacerbated when insulation is poorly fitted. This is often a consequence of the irregularity of existing structural components: when insulating an existing roof slope, for example, it is sometimes impossible to form a snug friction fit for rigid insulation because (a) rafter sections are often twisted to a certain extent and (b) rafters are seldom perfectly parallel. Gaps between the insulating material and structural components add to heat losses.

To minimise bridging via timber components and gaps in the insulation, it is practice to adopt a two-layer approach: insulation is fitted between studs, rafters and joists, with an additional layer fixed to the face of the same studs, rafters and joists (see Fig. 10.4b for an example).

Particular care should be taken when forming junctions between insulated surfaces, such as ceilings, roof slopes, stud walls and cavity walls. Insulation must be brought into close contact at junctions to maintain the continuity of the thermal envelope, so precise cutting and fixing are essential. Care should also be exercised at windows and other external openings, with insulation returned into reveals. Rigid insulating boards are available in thicknesses down to 12 mm to assist in the elimination of bridges in awkward corners and angles.

## INSULATION FOR WALL AND ROOF ELEMENTS

## Existing (retained) solid brick masonry walls

### U-value required: 0.30 W/m² K

As noted earlier in this chapter, solid brick masonry walls are relatively poor insulators. An unmodified 9" wall is assumed to have a U-value of about 2.1 W/m² K (Fig. 10.2a), or worse. Where a wall of this sort is to be incorporated within the thermal envelope of the conversion, it must be upgraded to conform to the 'improved' value of 0.30 W/m² K using one of the approaches outlined earlier in this chapter.

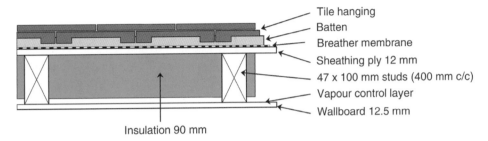

Tile hanging
Batten
Breather membrane
Sheathing ply 12 mm
47 x 100 mm studs (400 mm c/c)
Vapour control layer
Wallboard 12.5 mm

Insulation 90 mm

a) Insulation between studs
potential U-value: 0.31 W/m²K.
Note: cold bridging risk on studs.

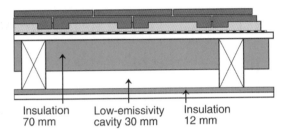

Insulation 70 mm    Low-emissivity cavity 30 mm    Insulation 12 mm

b) Insulation between studs and on internal face
potential U-value: 0.28 W/m²K.
Note: cold bridging minimised.

**Fig. 10.4**   New tile-hung stud wall: insulation with rigid PIR foam boards. Indicated U-values are for guidance only – insulant manufacturer's data should be consulted in all cases.

## New solid brick masonry walls

### *U-value required: 0.28 W/m² K*

New solid masonry walls may be constructed when an existing gable wall is extended to form a flank gable, or when a hip-to-gable conversion is undertaken. In these cases, the thickness of the new wall will generally be limited to that of the existing wall structure for practical reasons. The standard for new walls is 0.28 W/m² K, slightly better than that for existing walls.

As with existing walls, the most practical method is to provide insulating material inside the building (Figs 10.2b, c and d). Where both old and new elements are incorporated in a single wall structure (e.g. in a flank gable construction), there would be little to be gained by adopting two different standards of insulation on adjoining sections. Note that the thickness of insulation needed to achieve a U-value of 0.28 W/m² K is only marginally greater (about 5 mm) than that needed for 0.30 W/m² K.

## New solid blockwork walls

### *U-value required: 0.28 W/m² K*

New walls may be constructed using solid autoclaved aerated concrete (AAC) blockwork (Fig. 10.3). This approach may be adopted for buildings with either existing cavity or solid masonry walls. This would apply, for example, when a new gable wall of the same thickness as a flank wall beneath it is raised as part of a hip-to-gable conversion. Standard metric block thicknesses coordinate reasonably well with both 9" solid walls (215 mm blocks) and 'traditional' 10½" cavity configurations (265 mm blocks).

Either lightweight or ultra-lightweight aerated concrete blockwork may be used. Typically, lightweight blocks have a thermal conductivity of approximately 0.18 W/m K; ultra-lightweight blocks are more effective insulators with a conductivity value of about 0.11 W/m K.

The advantages of constructing walls in solid blockwork are speed and cost. The main disadvantage is that aerated concrete blockwork alone will not generally achieve the new-wall U-value (0.28 W/m²K) set out in ADL1B without the addition of supplementary insulation. AAC blockwork also requires an external weather-resisting finish.

However, it is still possible to achieve results that are significantly better than the overall limiting U-value for walls (0.70 W/m²K). The compressive strength of lightweight and ultra-lightweight blockwork should be considered where beam bearings are to be introduced.

## Existing (retained) cavity masonry walls

### *U-value required: 0.55 W/m²K*

In a dwelling with cavity walls, the cavity – generally – continues into the gable, although this is not universally the case: even in recently constructed dwellings, the cavity is sometimes closed at ceiling height on the top-floor level and the gable above constructed in solid masonry.

The overall thickness of pre-WWII cavity masonry walls is usually about 10½" (265 mm). Typically, the wall comprises an outer brick leaf, a cavity (usually a nominal 2"/50 mm) and an inner leaf in brick, clinker or concrete block. The thermal performance of such walls varies. With an inner leaf in medium-density concrete blockwork, an overall U-value of about 1.6 W/m²K would be typical. Better overall values could be expected when the inner leaf is constructed in clinker block (1.4 W/m²K).

If the cavity has been filled with insulating material (blown mineral wool or urea formaldehyde (UF) foam, for instance), the wall will provide a higher level of thermal performance. In these cases, a value of 0.50 W/m²K is commonly assumed.

In dwellings of the post-war period, aerated concrete blocks started to take the place of dense concrete, clinker and brick for inner-leaf construction. The combination of improved materials and wider cavities has led to considerable improvements in the thermal performance of external walls, with U-values in new-build down to approximately 0.40 W/m²K by the end of the 1970s.

Retro-fit cavity wall insulation was introduced in the late 1960s and cavity insulation for new-build in England and Wales became routine from the mid-1980s. However, even where cavity insulation has been provided, it should not be assumed that it extends to full gable height: it may be necessary to 'top up' cavity insulation to bring it up to the apex. If this approach is adopted, the 'top up' insulant should match the existing material. For practical reasons, this approach could only reasonably be applied with materials such as blown mineral wool, EPS beads or UF foam.

An alternative and sometimes less troublesome approach for existing cavity walls is to apply insulating material to the inner face of the wall. Fig. 10.5 illustrates some common cavity configurations.

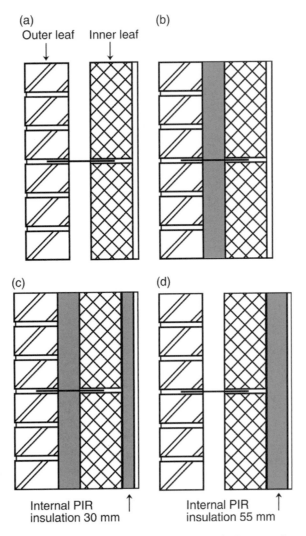

(a)
Outer leaf   Inner leaf

(b)

(c)
Internal PIR
insulation 30 mm

(d)
Internal PIR
insulation 55 mm

**Fig. 10.5** Existing cavity walls: 102.5 mm brick/50 mm cavity/100 mm medium-density concrete blockwork. Indicated U-values are for guidance only – insulant manufacturer's data should be consulted in all cases. (a) Wall cavity un-insulated assumed U-value: 1.6 W/m² K, (b) Wall cavity insulated (e.g. blown mineral wool) potential U-value: 0.50 W/m² K, (c) Wall cavity and internal face insulated potential U-value: 0.30 W/m² K, (d) Insulation of internal face only potential U-value: 0.30 W/m² K.

## New cavity masonry walls

### *U-value required: 0.28 W/m² K*

The structural thickness of the existing wall will determine the overall thickness of the new section and, in most cases, this will be 265 mm. It is possible to achieve a U-value of approximately 0.40 W/m² K with this thickness of structure by constructing the inner leaf from ultra-lightweight blockwork and providing cavity fill, such as mineral wool, as the

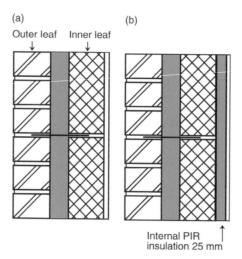

**Fig. 10.6**  New cavity walls: 102.5 mm brick/insulated 50 mm cavity/100 mm aerated concrete blockwork inner leaf (block conductivity: 0.11 W/m K). Indicated U-values are for guidance only – insulant manufacturer's data should be consulted in all cases. (a) Wall cavity insulated potential U-value: 0.40 W/m² K, (b) Wall cavity and internal face insulated potential U-value: 0.28 W/m² K.

wall is raised. However, in order to achieve the new-wall U-value of 0.28 W/m² K, some supplementary insulation to the inner face would normally be needed (Fig. 10.6).

## New tile hung stud walls

### *U-value required: 0.28 W/m² K*

New stud walls must be insulated to the same standard as new masonry walls. To achieve this, insulating material – generally PIR, PUR or phenolic foam boards – is friction-fitted between the studs.

Insulating between 100 mm thick studs alone, however, may not be enough to achieve the new-wall value of 0.28 W/m² K (Fig. 10.4a), so an additional internal layer of insulation is provided to limit thermal bridging (Fig. 10.4b). Note that the construction illustrated in Fig. 10.4a provides a lower standard of insulation than that shown in Fig. 10.4b, even though the overall thickness of insulating material is greater. At greater stud thicknesses (125 mm or more), bridging is reduced and insulation over internal stud faces may be omitted.

## Existing (retained) or new pitched roof

### *U-value required: 0.18 W/m² K*

The standard for both new and existing (retained) pitched roof slopes with insulation between the rafters is the same under guidance set out in ADL1B. Figs 10.7a and b illustrate methods of achieving a U-value of 0.18 W/m² K in a typical tiled roof with 47 × 100 mm

(a)

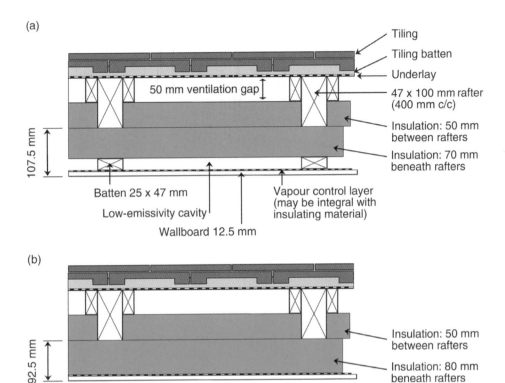

Tiling
Tiling batten
Underlay
47 x 100 mm rafter (400 mm c/c)
Insulation: 50 mm between rafters
Insulation: 70 mm beneath rafters

50 mm ventilation gap

107.5 mm

Batten 25 x 47 mm
Low-emissivity cavity
Wallboard 12.5 mm
Vapour control layer (may be integral with insulating material)

(b)

92.5 mm

Insulation: 50 mm between rafters
Insulation: 80 mm beneath rafters

**Fig. 10.7**   Pitched roof insulation with rigid PIR foam boards. Indicated U-values are for guidance only – insulant manufacturer's data should be consulted in all cases. (a) Insulation between and below rafters; plasterboard on battens potential U-value: 0.18 W/m²K, (b) Insulation between and below rafters; battens omitted potential U-value: 0.18 W/m²K.

rafters at 400 mm centres. The method illustrated in Fig. 10.7b offers the 'thinnest' construction. In both cases, the additional loading created by insulation, internal battens and wallboard must be considered.

## New flat warm roof

### *U-value required: 0.18 W/m² K*

Fig. 10.8a illustrates one method of achieving a U-value of 0.18 W/m²K with a warm deck flat roof using rigid PIR foam boards. Because the insulation is not bridged, the approach illustrated will serve for any flat roof, irrespective of joist dimensions. Note that the upper surface of the insulation must be suitably adapted to withstand the application of hot-bedded roofing where a built-up finish is to be applied.

Although relatively easy to construct (ventilation is not required), warm deck roofs are considerably thicker than cold decks and this sometimes causes difficulties. In the case of a warm flat roof spanning 3.5 m, the additional thickness of firring, insulation and final roof covering *above* the flat roof joists is likely to exceed 200 mm at the highest point. Adopting a warm flat roof therefore requires careful planning in cases where the height of

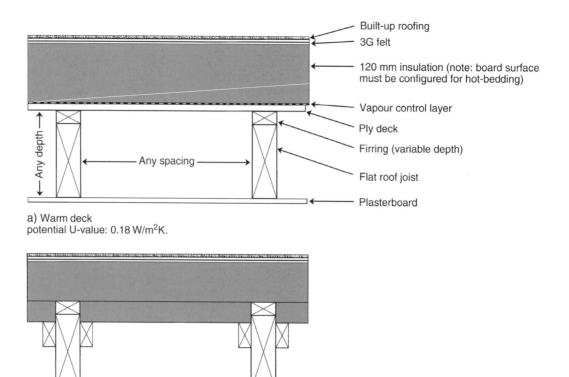

a) Warm deck
potential U-value: 0.18 W/m²K.

b) Warm deck (hybrid)
potential U-value: 0.18 W/m²K.
Note: This configuration is not universally accepted.

**Fig. 10.8**    Flat roof (warm deck – unventilated) with rigid PIR foam board insulation. Indicated U-values are for guidance only – insulant manufacturer's data should be consulted in all cases.

the finished roof is critical (see also Fig. 9.10). A hybrid warm roof, with insulation over and between joists, is accepted by some building control bodies (Fig. 10.8b).

## New flat cold deck

### *U-value required: 0.18 W/m² K*

Fig. 10.9 illustrates four cold-deck configurations achieving a U-value of 0.18 W/m²K with common joist sections. Note that it is necessary to maintain a 50 mm ventilation gap above the insulation material in all cold roof structures – the depth of firring should therefore be discounted unless it is greater than 50 mm at its thinnest end.

## WINDOWS AND OTHER OPENINGS

Windows, roof windows and doors are described as *controlled fittings* in ADL1B. Windows and roof windows should have either a Window Energy Rating (WER) of Band 'C' or better, or a U-value of 1.6 W/m²K or better (see Table 10.4), while external doors must

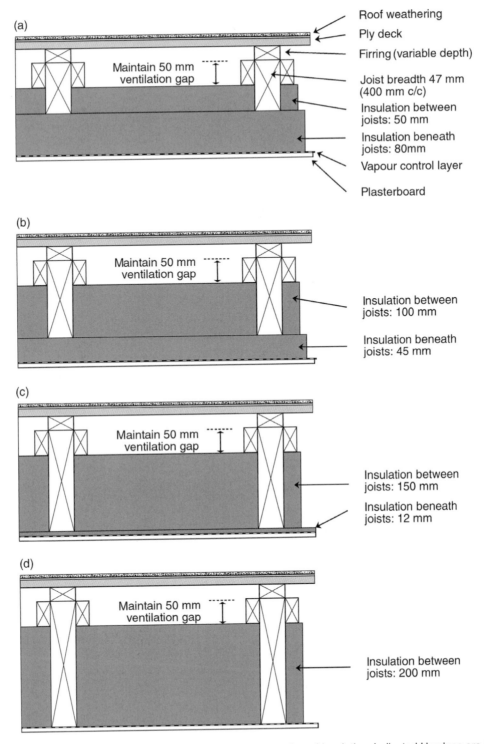

**Fig. 10.9** Flat roof (cold deck – ventilated) with rigid PIR foam board insulation. Indicated U-values are for guidance only – insulant manufacturer's data should be consulted in all cases. (a) Cold deck – joist depth 100 mm potential U-value: 0.18 W/m² K, (b) Cold deck – joist depth 150 mm potential U-value: 0.18 W/m² K, (c) Cold deck – joist depth 200 mm potential U-value: 0.18 W/m² K, (d) Cold deck – joist depth 250 mm potential U-value: 0.18 W/m² K.

**Table 10.4**   Standards for controlled fittings.

| Fitting | Standard |
| --- | --- |
| Window, roof window or rooflight | WER Band 'C' or better, or U-value: 1.6 W/m² K |
| Doors with more than 50% of the internal face glazed | U-value: 1.8 W/m² K |
| Other doors | U-value: 1.8 W/m² K |

Where test data and calculated values are not available, reference may be made to *SAP 2009*, Table 6e (Default U-values (W/m² K) for windows, doors and roof windows).

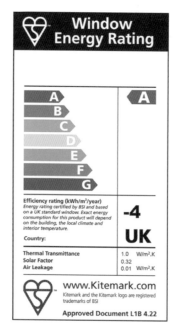

**Fig. 10.10**   Window Energy Rating (WER) product certification label. For a colour version of this figure, please see the colour plate section.
Courtesy BSI.

achieve a U-value of 1.8 W/m² K or better. The U-value applies to the *entire* window or door assembly and includes both the frame and glazing.

Third-party certification – such as the WER scheme – provides assurance that products achieve a declared level of performance (see Fig. 10.10). In some cases, the manufacturer will provide only a declared U-value for the fitting.

In the absence of test data or calculated values for windows and external doors, reference can be made to SAP 2009, Table 6e, which provides default U-values for windows, doors and roof windows. These are based on notional windows and the values set out are conservative, with only a handful of triple-glazed assemblies meeting or exceeding the thermal efficiency requirements.

Values for roof windows in ADL1B are based on the U-value having been assessed with the roof window in a vertical position. If a window unit has been assessed in a plane other than the vertical, the standards set out in the Approved Document should be modified by making a slope-dependent adjustment set out in BR 443 *Conventions for U-values*, BRE 2006.

It should be noted that loft conversions are prone to overheating in hot summer weather. Roof windows positioned at a relatively high point on the roof slope will provide some assistance in controlling the build-up of heat. By contrast, low-level windows are less effective in aiding the dispersion of hot air that accumulates at apex or ceiling height (see also *Purge ventilation* below).

## Area of windows

Windows, roof windows, rooflights and external doors are relatively poor insulators. In order to limit heat loss through them, the guidance in ADL1B is that openings of this sort be restricted to 25% of floor area of the extension. To this proportion may be added the area of any windows or doors that, as a result of the work, no longer exist or are no longer exposed (the latter guidance is generally only relevant when a building is extended outwards at a lower level).

Window areas in excess of 25% may be acceptable if either (a) the U-value of the fitting is improved or (b) that compensatory measures are applied. These may be based either on the area-weighted U-value or whole-building calculation method described at the beginning of this chapter.

The Approved Document cautions against window areas *less* than 20% of the total floor area because this might result in poor daylighting levels and a concomitant increase in the need for electric lighting.

Window, roof window, rooflight and door units should be draught proofed and have a performance that is no worse than that set out in Table 10.4. Insulated cavity closers should be installed where appropriate.

In Table 10.4, the U-value has been assessed with roof windows in the vertical position.

## RISKS ASSOCIATED WITH INSULATION

There are technical risks associated with the provision of insulation, particularly where the insulating material is applied to the inside of an existing building. This is primarily because high levels of insulation make elements outside the insulated envelope much colder and thus more susceptible to the damaging effects of condensation. Effective vapour control (i.e. the provision of uninterrupted VCLs and appropriate ventilation) may limit inside-to-outside vapour movements but, in the case of alterations to existing buildings, such as loft conversions, it is not usually possible to eliminate unwanted moisture movements entirely.

## Surface condensation

A building with insulation fixed to internal surfaces warms up relatively rapidly but because insulating materials (unlike solid masonry) have a low thermal mass, cooling is correspondingly swift when heating is turned off. This increases the risk of surface condensation occurring. Adequate ventilation will serve to limit the formation of surface condensation. Guidance is set out in Approved Document F *Means of ventilation*.

## Interstitial condensation – all elements

Vapour-rich air that escapes from the insulated building envelope will condense in cold structural voids, particularly ones that are not adequately ventilated. Condensation of this sort – interstitial condensation – is potentially destructive over extended periods of time, particularly where it occurs on timber structural members or within insulating materials. Ensuring continuity of the VCL will minimise the amount of vapour that escapes from the building; effective ventilation of cold roof structures (see also Chapter 9) will limit the accumulation and condensation of any water vapour that does escape.

## Spalling risk – masonry walls

External walls that are insulated from the inside are no longer warmed by the building and are therefore at greater risk of freeze/thaw action. Bricks and mortar may spall if moisture is present over extended periods of time.

## ELECTRIC LIGHTING

Fixed internal lighting is described as a *fixed building service* in the Building Regulations and ADL1B. However, detailed guidance on lighting provision is no longer included in the Approved Document. Instead, users are referred to the *Domestic Building Services Compliance Guide (2010)*. This is published by NBS on behalf of the Department for Communities and Local Government and has similar status to an Approved Document.

The *Domestic Building Services Compliance Guide* sets out the following recommended minimum standards for new and replacement fixed internal lighting:

(a)  In areas affected by the building work, provide low energy light fittings (fixed lights or lighting units) that number not less than three per four of all the light fittings in the main dwelling spaces of those areas (excluding infrequently accessed spaces used for storage, such as cupboards and wardrobes).
(b)  Low energy light fittings should have lamps with a luminous efficacy greater than 45 lamp lumens per circuit-watt and a total output greater than 400 lamp lumens.
(c)  Light fittings whose supplied power is less than 5 circuit-watts are excluded from the overall count of the total number of light fittings.
Light fittings may be either:
- Dedicated fittings which will have separate control gear and will take only low energy lamps (e.g. pin-based fluorescent or compact fluorescent lamps); or
- Standard fittings supplied with low energy lamps with integrated control gear (e.g. bayonet or Edison screwbase compact fluorescent lamps).
Light fittings with GLS tungsten filament lamps or tungsten halogen lamps would not meet the standard.

## Practical implications

The current guidance represents a slightly more pragmatic approach to lighting provision, at least as far as light fittings are concerned: in earlier guidance (ADL1B 2006), the emphasis was on providing light fittings that could accommodate *only* energy-efficient lamps. The current guidance, however, indicates that 'standard' bayonet cap (BC) and Edison screw (ES) fittings are acceptable, provided that low energy lamps – such as compact fluorescent lamps – are used in them.

The inclusion of 'standard' fittings means that the use of mains voltage GU10 downlight fittings could also reasonably be considered, provided compact fluorescent lamps (CFLs) are used in three out of every four fittings. This means it is possible to provide energy-efficient general lighting with fittings that are flush with the ceiling. In loft conversions, recessed light fittings are of particular value because headroom is often limited.

Note that GU10 CFL lamps are slightly larger than incandescent versions and are not dimmable. Purpose-made GU10 CFL fittings may only be able to safely accommodate lamps rated at a maximum of 11 W. Recessed light fittings in the ceiling of a loft conversion would not normally need to be fire rated (but ceiling penetrations on lower floors caused by downlight fittings would need to be made fire resisting).

Wall lights fitted with standard ES or BC compact fluorescent lamps offer an effective alternative to recessed ceiling lighting where headroom is limited. Wall-mounted fittings remove the need to cut into the ceiling, thus maintaining the integrity of roof insulation and vapour control layers, as well as preserving airtightness.

With 'standard' fittings capable of accommodating both CFL and incandescent lamps (such as BC, ES or GU10 light fittings), the long-term energy efficiency of the lighting system will depend on the continuing use of the correct type of lamp. It would therefore be reasonable to provide householders with information to this effect. The Building Regulations requirement on providing information is set out in regulation 40.

## HEATING AND HOT WATER SYSTEMS

Detailed guidance on new and replacement heating and hot water systems is provided in the *Domestic Building Services Compliance Guide* (2010), rather than the Approved Document. Heating and hot water systems are classified as *fixed building services*. If the heating system of the dwelling is completely replaced, the building control body will require evidence of commissioning.

In the majority of cases, a loft conversion will not trigger the need for an entirely new or replacement system. Additional radiators and hot water for a bathroom, for example, will be provided by the simple extension of existing heating and hot water systems. However, the current guidance does not provide information on the obligations associated with extensions to existing systems.

When additional radiators are added to an existing central heating system and the hot water system extended, the following approach would be reasonable:

- All new radiators should be equipped with individual radiator controls, such as thermostatic radiator valves (TRVs) except reference rooms with a thermostat, and bathrooms.
- If the loft conversion constitutes a zone, a room thermostat or programmable room thermostat should be provided.
- Pipework should be insulated (in cold areas of the building) and labelled as appropriate.
- The installer should provide the owner of the dwelling with information on the energy-efficient operation of the extended system.

## Providing information about energy efficiency

Regulation 40(2) of the Building Regulations 2010 states that where paragraph L1 of Schedule 1 (conservation of fuel and power) imposes a requirement in relation to building work:

> The person carrying out the work shall not later than five days after the work has been completed provide to the owner sufficient information about the building, the fixed building services and their maintenance requirements so that the building can be operated in such a manner as to use no more fuel and power than is reasonable in the circumstances.

In cases where a new heating system is installed, for example, compliance with the requirement would be demonstrated by providing the owner of the dwelling with operating and maintenance instructions aimed at achieving economy in the use of fuel and power. The Approved Document (Section 7) indicates that these should be:

(1) Written in terms that householders can understand
(2) Durable
(3) Directly related to the particular system or systems installed

Instructions (in the case of a new heating system) should include information on:

- Making adjustments to timing, temperature and flow control settings
- Routine maintenance needed to keep the system operating efficiently

## LOFT INSULATION WHEN A LOFT IS NOT CONVERTED

The target U-value for loft insulation is $0.16\,W/m^2\,K$. The current guidance is to provide loft insulation as a quilt to a thickness of 250 mm using mineral or cellulose fibre laid between and across ceiling joists. Loose fill insulant or rigid insulation boards may also be used to achieve the target value.

The condensation risk in the roof space must be assessed and provision for ventilation to limit condensation made where necessary. The need to provide access to services – and insulation for them – must also be taken into account.

By virtue of Regulation 12(6)(b), the installation of thermal insulation in a roof space or loft space need not be notified to building control when *only* this work is carried out, and the work is not carried out to comply with any requirement in the Building Regulations.

## VENTILATION FOR OCCUPANTS

The requirement to provide adequate means of ventilation for people in buildings is set out in Part F of the Building Regulations 2010. The requirement is now supported by three sets of official guidance. These are:

- *Approved Document F – Ventilation* (2010)
- *Domestic Ventilation Compliance Guide* (2010)
- *Domestic Building Services Compliance Guide* (2010)

Note that ventilation in the context of Part F refers *only* to the provision of fresh air and removal of stale air for the benefit of people in the building. The ventilation of building fabric, which includes ventilation of roof structures to prevent condensation, is dealt with separately in Part C of the Building Regulations with guidance in Approved Document C *Site preparation and resistance to contaminants and moisture* (2004) (see Chapter 9). Air supply for combustion appliances is also considered separately, with reference to Part J of the Building Regulations and specific guidance in Approved Document J *Combustion appliances and fuel storage systems* (2010).

The balance between the requirement to provide ventilation for people and the need for thermal efficiency is a delicate one. In order to be energy efficient, buildings need to be airtight. Both Approved Documents L and F emphasise the need to reduce unwanted air leakage through new parts of the building envelope. But in order to be safe and comfortable, occupants need fresh air.

Because hermetically sealed dwellings are not desirable, the official guidance in Approved Document F and elsewhere strikes a balance between the well-being of occupants and the requirement to limit heat losses and gains caused by infiltration (uncontrolled air movements between the inside and outside of buildings). Approved Document F thus focuses on purpose-provided ventilation to facilitate controllable air exchange.

In a relatively simple extension to a dwelling, such as a loft conversion, adequate ventilation would normally be achieved through combinations of the following three methods.

## Background ventilation

This is nominally continuous low-rate ventilation of rooms or spaces. It is also described in the Approved Document as 'whole-building ventilation', 'whole dwelling ventilation' and 'general ventilation'.

This may be provided by manually controlled trickle ventilators in windows, but note that these are not routinely provided by window manufacturers. In high-performance windows, for example, trickle ventilators may be intentionally omitted in order to achieve airtightness targets.

Wall ventilators may be used instead of trickle vents to provide background ventilation. Both wall and window vents should be positioned at about 1.7 m above floor level to limit cold draughts. Ventilators of this sort are intended to be left in the 'open' position in occupied rooms.

Note that 'equivalent area' instead of 'free area' is now used for sizing background ventilators, including trickle ventilators. The method for measuring equivalent area (which measures the airflow performance of the ventilator) is set out in BS EN 13141-1:2004. The earlier 'free area' (geometric measure) of a trickle ventilator is approximately 25% larger than the new 'equivalent area'.

## Purge (rapid) ventilation

This is the technical term for manually controlled high-rate ventilation. Purge ventilation is usually provided by openable windows or an external door, but may be provided by a fan instead.

Purge ventilation provisions are based on the relationship between the opening area of windows and external doors, and the floor area of the room being ventilated. Note that for the purposes of fire safety, the minimum size requirements for means of escape windows must be observed in all cases (see Chapter 4).

- *Parallel sliding window such as a vertical sliding sash window*
  Area of the opening part should be at least 1/20th of the room's floor area.
- *Hinged or pivot window opening 30° or more*
  Area of the opening part should be at least 1/20th of the room's floor area.
- *External door*
  Area of the opening part should be at least 1/20th of the room's floor area.
- *Hinged or pivot window opening between 15° and 30°*
  Area of the opening part should be at least 1/10th of the room's floor area.
- *Hinged or pivot window opening less than 15°*
  Not suitable for purge ventilation; alternative measures required.

In a room containing *more than one openable window*, the areas of all the opening parts can be summed to achieve the required proportion of the floor area. This proportion is determined by the opening angle of the largest window.

In a room containing *more than one external door*, the opening areas may be summed to achieve at least 1/20th of the room's floor area.

In a room containing *a combination of at least one external door and at least one opening window*, the areas of all the opening parts may be added to achieve at least 1/20th of the room's floor area.

## Extract ventilation

This is the direct removal of air from a space or spaces inside the building to outside. Extract ventilation may be mechanical (e.g. fan) or natural (e.g. passive stack ventilation). Mechanical ventilation is categorised as a *fixed building service* (see Glossary). The *Domestic Building Services Compliance Guide* (2010) indicates that the specific fan power (SFP) of an intermittent extract ventilation system should not be worse than 0.5 W/(l/s).

Further guidance is contained in the *Domestic Ventilation Compliance Guide* (2010). The key points are:

- Holes/ductwork that pass through an external wall should have a slight downward angle to prevent water ingress.
- Where possible, rigid ducting should be used in preference to flexible ductwork.
- A condensate trap should be provided in vertical ducting to prevent the backflow of moisture.

## VENTILATION – PRACTICAL MEASURES

## All rooms

Provide an undercut with a minimum area of 7600 mm² to internal doors to promote air transfer throughout the dwelling. In the case of a standard 760 mm width door, this would be an undercut of 10 mm above the final floor finish.

# Habitable room (with external wall)

A habitable room includes a bedroom (but see Glossary for full definition). The following provisions would be reasonable:

- *Background ventilation*: provide ventilator(s) with an equivalent area of at least 5000 mm² (e.g. window trickle vents).
- *Purge ventilation*: provide openable window(s) or external door(s) of appropriate dimensions.
- *Extract ventilation*: not normally required.

# Habitable room (with *no* external wall)

As noted above, a habitable room includes a bedroom (see Glossary for full definition). An internal habitable room with no openable windows may be ventilated through an appropriately ventilated adjoining habitable room. The two rooms are treated as a single room for ventilation purposes and there must be a permanent opening between them. The following provisions would be reasonable:

- *Background ventilation*: in external room, provide ventilator(s) with an equivalent area of at least 8000 mm².
- *Purge ventilation*: in external room, provide openable window(s) or door(s) scaled to the total combined floor areas of both rooms.
- *Extract ventilation*: not normally required.

Note that an 'internal room' (Approved Document F) is also likely to constitute an 'inner room' for the purposes of Approved Document B, *Fire safety*. In the case of a loft conversion, a habitable inner room is unlikely to be acceptable.

# Bathroom (with external wall)

For the purposes of ventilation, a bathroom can include anything from a simple shower room to a fully equipped bathroom incorporating a shower, bath, WC and hand washbasin. In Approved Document F, a bathroom is considered to be a wet room. The following provisions would be reasonable where the bathroom has an external wall:

- *Background ventilation*: provide ventilator(s) with an equivalent area of 2500 mm².
- *Purge ventilation*: provide an openable window (no minimum size)
- *Extract ventilation*: Minimum intermittent extract rate of 15 litres/s.

Intermittent extract ventilation may be operated manually and/or by a sensor (e.g. humidity, occupancy/usage, pollutant release). Intermittent extract fans should be installed as high as practical and preferably less than 400 mm below ceiling level. Extract fans and background ventilators should be at least 500 mm apart.

# Bathroom (with *no* external wall)

As noted above, a bathroom (for the purposes of ventilation) can include everything from a simple shower room to a fully equipped bathroom with a shower, bath, WC and hand washbasin. In Approved Document F, a bathroom is considered to be a wet room.

- *Background ventilation*: see *Extract ventilation* (below)
- *Purge ventilation*: see *Extract ventilation* (below)
- *Extract ventilation*: Minimum intermittent extract rate of 15 litres/s.

In rooms with no openable window, the intermittent extract fan should have a 15-minute overrun. In rooms with no natural light, the fans could be controlled by the operation of the main room light switch. Intermittent extract fans should be installed as high as practical and preferably less than 400 mm below ceiling level.

## WC (with external wall)

For the purposes of Part G, a room containing a WC must have a hand washbasin; for the purposes of Part F, a WC (sanitary accommodation) is considered to be a wet room.

- *Background ventilation*: provide ventilator (e.g. window trickle ventilator) with an equivalent area of at least 2500 mm$^2$.
- *Purge ventilation*: provide openable window(s) of appropriate dimensions.
- *Extract ventilation*: not required if purge window is scaled as above.

## WC (with *no* external wall)

For the purposes of Part G, a room containing a WC must have a hand washbasin; for the purposes of Part F, a WC (sanitary accommodation) is considered to be a wet room. Humidity-controlled extract ventilation is not considered to be appropriate for WCs where odour, rather than water vapour, is the dominant pollutant.

- *Background ventilation*: see *Extract ventilation* (below).
- *Purge ventilation*: see *Extract ventilation* (below).
- *Extract ventilation*: Minimum intermittent extract rate of 6 litres/s.

In rooms such as a WC with no openable window, the intermittent extract fan should have a 15-minute overrun. In rooms with no natural light, the fans could be controlled by the operation of the main room light switch. Intermittent extract fans should be installed as high as practical and preferably less than 400 mm below ceiling level.

## Providing information about ventilation

Regulation 39(2) of the Building Regulations 2010 states that where paragraph F1(1) of Schedule 1 (ventilation) imposes a requirement in relation to building work:

> The person carrying out the work shall not later than five days after the work has been completed give sufficient information to the owner about the building's ventilation system and its maintenance requirements so that the ventilation system can be operated in such a manner as to provide adequate means of ventilation.

In the case of a loft conversion, the information requirement might be limited to providing the owner with the instruction manual for a mechanical extractor fan in a bathroom. But it might usefully be extended to an explanation of the need for background ventilation. Anecdotal evidence suggests that many householders would struggle to identify a trickle ventilator.

# 11 Lofts in context

Since 1990, loft conversions have increased the capacity of the UK housing stock by more than 200 million square feet. This increase in the amount of habitable accommodation, equivalent to a city the size of Manchester, has been achieved entirely within the footprint of the existing housing stock and without the loss of a single square foot of land.

This chapter considers some of the reasons householders choose to convert lofts, and why the conversion market has flourished in the UK. It also looks at the broader historical, social and environmental factors that influence the loft conversion market, including the growing interest in sustainability and energy efficiency.

## WHY CONVERT?

Loft conversions have grown in popularity for a number of reasons. The main one is that conversion allows owner-occupiers to enjoy the benefits of a larger house – typically, one with an extra bedroom and an additional bathroom – without the high costs, uncertainty and inconvenience associated with selling up and moving away. An additional advantage is that garden space, which is at a premium in urban areas, is preserved by extending upwards rather than outwards.

Depending on the local property market, adding an extra bedroom is one of the most cost-effective home improvements that householders can make. This has tended to remain the case even in times of relative instability in the housing market.

Typically, a loft conversion is less disruptive, and less expensive, than a ground-level extension because no foundations and little 'wet' construction is involved. A significant proportion of the building work can be carried out without builders needing to set foot in habitable parts of the existing dwelling. Work is carried out relatively swiftly, with most conversions completed in 4–6 weeks.

High property prices continue to act as a strong incentive to improve rather than move. But the higher capital cost of moving to a larger property is not the only reason people choose to stay put: in cities such as London, the fixed costs associated with moving – such as stamp duty land tax, removal charges, estate agents' and legal fees – may exceed the cost of a loft conversion. A significant number of conversions pay for themselves in this respect.

As well as saving money, extending an existing dwelling instead of moving house means families are able to protect hard-won social capital – places at local schools, networks of friends and local knowledge – which might otherwise be lost. Social capital defies valuation in the conventional sense, but it can take years to accumulate and cannot easily be replenished.

## LOFT CONVERSION STATISTICS

Official statistics suggest that the total number of 'pure' loft conversions (those converted since original construction, excluding as-built attics) was 1 076 000 or 4.8% of all dwellings in England in 2008. The proportion of conversions in pre-1919 dwellings is considerably higher, with lofts converted in 9.6% of the stock or about 455 000 dwellings (data from the *English Housing Survey 2008*, courtesy of the Economic and Social Data Service).

The number of dwellings with habitable roof spaces including as-built attics is greater still. Of the 22.2 million dwellings in England in 2008, 2.3 million dwellings – just over 10% of the housing stock – had an attic comprising a room or rooms with a staircase, permanent floor and natural lighting. The figure of 2.3 million includes both as-built rooms in the roof and loft conversions.

In Wales, the overall proportion of converted properties is slightly smaller. In 2008, 3.9% of all Welsh dwellings, approximately 51 000, had loft conversions (figures courtesy of the Welsh Assembly Government).

## UNDERLYING TRENDS

Loft conversions are not unique to Britain: France has its *aménagement de combles* and Germany its *Dachausbau*, for example. But the spectacular growth of loft conversion in the UK over the last 20 years – affirmed by the sheer number of conversions and the size and relative maturity of the specialist loft conversion sector – is significant by international standards.

The British appetite for loft conversions is a reflection of a number of social and economic trends. Aside from the need for an extra bedroom, British householders now expect additional bathrooms and WC facilities in even modestly sized houses; for practical purposes, it is easier – and more in line with conventional taste – to provide these in an upper storey rather than at ground level.

Changing work patterns are also a spur to loft conversions and an increasing number of conversions are now built specifically to provide a study or home office accommodation. This reflects a wider societal shift towards telecommuting and it is a trend that is likely to continue: currently, the UK has around 1.3 million home workers, with 3.7 million employees working from home occasionally, or who use home as a base.

## THE NATURE OF THE HOUSING STOCK

Social trends aside, additional reasons for the prevalence of loft conversions can be found in both the physical characteristics of the UK housing stock and patterns of ownership.

British housing is amongst the oldest in the developed world: in England, for example, 38% of the housing stock was constructed before 1945. Many of these older buildings are particularly well-adapted to conversion, with masonry walls capable of providing structural support and robustly constructed pitched timber roofs that are – generally – tolerant of significant modification.

Britain's housing stock is old – and getting older – for a number of reasons. One is that house building is at a historically low level. A contributory factor is that a high proportion of the existing stock – now more than two thirds – is in private ownership, so large-scale residential redevelopment of the sort seen in the post-war years would not be practical, even if the political appetite and funding were available to make it possible.

Piecemeal replacement of the stock also presents intractable problems. The preponderance of terraced and semi-detached housing in urban areas (Britain has the second-highest proportion of such housing in the EU-27) means that scope for an individual owner to rebuild a house from scratch is limited.

Rates of demolition in the UK are consequently exceptionally low. At current levels, more than half of today's housing stock will still be standing in 500 years' time. The distaste for demolition and the cachet associated with owning a 'period' home that has grown since the 1970s means that the UK, more than almost any other developed country, remains compelled to find the best ways it can to make to use the buildings it already has.

## PRACTICAL SUSTAINABILITY

Progressive tightening of the Building Regulations over the last 15 years has transformed the energy efficiency of both new buildings and extensions to existing ones. But as the emphasis on statutory sustainability has grown, so, too, has public interest in cutting emissions and reducing energy use. Loft conversions present a number of opportunities for enhancements to domestic efficiency.

## Renewable energy

A loft conversion provides an opportunity to install roof-mounted microgeneration technologies such as wind turbines, photovoltaic (PV) panels and solar hot water collectors.

An increasing number of householders are taking the opportunity to install such systems while the scaffolding is still up – although the orientation of the dwelling must be carefully considered in all cases. PV and solar hot water systems depend on sunlight and they work most effectively on south-facing roof slopes; where such a slope is replaced by a box dormer (collectors would then need to be mounted vertically), systems of this sort operate less efficiently.

### Solar photovoltaic

These generate d.c. electricity from sunlight using PV cells (Fig. 11.1). Output is converted to a.c. using an inverter and the electricity generated is then either consumed by the householder or – if it is surplus to needs – exported to the grid. According to the Energy

**Fig. 11.1**  Solar photovoltaic (PV) installation.

Saving Trust, a typical home PV system can produce around 50% of the electricity a household needs in a year.

### Solar thermal (ST)

These systems produce hot water using energy from the sun. In solar water heating systems, water (treated with anti-freeze) is circulated from roof-mounted collectors, where it is heated, to a heat exchanger within the hot water cylinder; water for domestic use is thus heated indirectly. A twin-coil hot water cylinder is required, with one coil for the solar collector loop and the other for heat exchange via a conventional boiler (which will still be required during the colder months). A small amount of electricity is consumed via the ST system's pump and control system. ST systems have the potential to provide most of the hot water required for domestic purposes during the summer months.

### Wind turbines

These produce electricity via turbines with an integral generator. Roof-mounted turbines (as opposed to mast-mounted ones) have the potential to generate between 1 and 2 kW. However, to approach optimum efficiency, most turbines require an average wind speed of 5 m/s or more – higher than the average wind speed in most inland urban areas. Planning considerations and turbulence caused by surrounding buildings also mean that this technology is not always well suited to use in cities.

## Reducing solar gain

Loft overheating is a common householder complaint. To reduce overheating in the summer months – and the subsequent introduction of air conditioning – the orientation of the conversion and window positioning should be carefully considered at the design stage. While insulated walls and roofs will reduce the rate at which inside air temperature rises, extensive glazed areas – particularly south and west-facing ones – will cause the roof space to heat up rapidly.

## Green roofs

Green or living roofs have grown in popularity in recent years. These have a number of practical benefits: as well as providing natural insulation in winter, they help to regulate internal temperatures in summer, reducing heat gain and promoting cooling through evaporation. The relatively high mass of the green roof also provides enhanced sound insulation.

There are aesthetic benefits too (although clearly none if the green roof forms the top of a box dormer, in which case it is unlikely to be visible). Green roofs – which generally incorporate resilient sedum plants – also promote biodiversity and provide a valuable habitat for insects and flora that would otherwise not be available. They also reduce run-off, slowing down the rate at which rainwater is released. This reduces loading on drainage networks, an increasing concern in urban areas.

The drawbacks of green roofs include the need for a stronger roof structure and the cyclical stresses caused by repeated wetting and drying of the roof: on a typical box dormer, a heavy downpour could increase the loading on the roof by more than a tonne in a matter of minutes. In addition, the edges of the roof must be configured to prevent wind uplift of the sedum matting, so a parapet or suitable raised protective edging must be provided.

## Water conservation

Water is an increasingly scarce resource in parts of the UK. The problem is becoming acute in the south-east, where the combination of population growth and increasing per capita consumption means that the region has less water available per person than either Sudan or Syria.

The selection of water-efficient fittings can play a significant part in reducing water consumption. Lower-flow taps and showerheads also help to cut energy bills because demand for hot water is reduced.

### Water-efficient WCs

New WCs are required to have a flush not exceeding 6 litres. However, further water savings can be achieved by specifying low flush WCs. These include siphon WCs providing a 4 litre flush, and dual-flush valve WCs offering a combined 4 and 2 litre (4/2 litre) action.

### Flow-restrictor valves

These achieve savings by regulating the flow of water. Restrictor valves for showers offer flows as low as 6 litres/minute (compared to 20 litres for a typical mains shower). Restrictor valves are not suitable for electric showers. Flow-regulating valves are also available for basins.

### Taps and showerheads

In taps, spray insets and aerators (which mix air with water) help to reduce consumption by altering the characteristics of water as it leaves the tap. Aerated showerheads work in a similar way – effectively by adding bubbles – and can reduce water use by up to 60%. As with flow-restrictor valves (above), aerated showerheads should not be used in conjunction with electric showers.

### Reduced-capacity baths

These are simply baths that require less water to fill, and are marketed as 'water-saving' or 'low-volume' baths: internal contouring means these offer a deep bath, without excessive use of water. Typically, these are configured for a 140 litre fill, rather than the 230 litres needed to achieve a similar depth with a conventional bath.

### Recycling water

Making use of second-hand water – such as rainwater from roofs and greywater from basins and baths – has certain environmental benefits. However, the costs and complexities associated with greywater recycling and rainwater harvesting need to be carefully considered.

#### Rainwater harvesting

Rainwater run-off from roofs is collected via guttering and stored in a tank below ground. This water may then be pumped back for flushing lavatories and watering gardens. In some cases, harvested water is used in washing machines. Running costs – which include routine maintenance and energy costs for pumping – should be taken into account. A rainwater butt is a low-tech but effective alternative.

#### Greywater recycling

This involves collecting non-foul domestic wastewater streams, including waste from baths, showers and basins. This water must be treated before re-use. Generally, greywater is re-used to water gardens, but in some cases, it may be used to flush WCs. The need to provide tank storage, treatment, distribution, pumping and regular maintenance must be considered.

## Reducing construction waste and re-using materials

Almost 10 million tonnes of waste wood are produced in the UK every year, most of which ends up as landfill according to Defra, the Department for Environment, Food and Rural Affairs. Currently, only around 15% of this is recycled. Loft conversions alone are responsible for producing between 5000 and 10000 tonnes of waste timber every year.

Timber is just the tip of the iceberg. The building industry is one of the largest producers of waste in the UK, and as well as timber, substantial quantities of slates, tiles, fixings, sheet materials, bricks and blocks are regularly consigned to the scrapheap, contributing to the estimated 120 million tonnes of construction-related waste generated annually.

Loft conversions and other types of refurbishment generate waste streams that comprise both new and original materials. With new materials such as timber, plasterboard and insulation, a degree of cutting wastage is inevitable. But a surprising amount of 'virgin' material finds its way onto skips as well, simply because too much was ordered in the first place. This represents a waste of both money and resources, and in many cases, it can be avoided: an accurate survey of the existing building at the design phase and careful study of drawings when ordering materials can yield savings.

Existing materials present a slightly different challenge. Recovering them intact, and then processing them, can be a time-consuming exercise. Some materials are, of course, more easily recovered than others. Re-using existing slates and tiles, for example, is relatively straightforward, provided fixing holes and nibs are intact. As noted in Chapter 9, these may provide a well-matched and cost-free source of materials.

Mineral wool loft insulation laid at rafter level, if it has been recently installed, may also be re-used, providing it is of adequate thickness and is in good condition. Note that flattened and dirty insulation will not be satisfactory for meeting the requirements of parts B and E of the Building Regulations.

Brick is now routinely recycled. Providing it was originally bonded with a lime-based mortar, cleaning up bricks for re-use can be easily achieved with hand tools.

Redundant rafters and floor joists have always provided a ready source of material for minor applications, such as floor, wall and roof noggings. But systematic recovery and recycling of longer timbers is a slightly more risky proposition. Certain existing timbers – ceiling joists, for example – are likely to be riddled with nails where they have supported a traditional plaster and lath ceiling and are therefore not good candidates for re-use. Rafters are rather more promising, provided that batten nailing and other fixings are carefully removed. One potential application is in the construction of non-load-bearing internal partitions, although any timber is capable of re-use for structural purposes provided it has been visually strength graded (see Chapter 5).

When re-using timber, it is essential that safety precautions are carefully observed. These might include the wearing of gloves, eye protection and the use of hand saws rather than power saws for cutting where there is a risk of encountering embedded nails and fixings.

Further information on the re-use of building materials can be found in the *Reclaimed Building Products Guide*, published by Waste & Resources Action Programme (WRAP).

## ENERGY PERFORMANCE CERTIFICATES (EPCs) AND INSULATING TO A HIGHER STANDARD

The standards for 'new' and 'retained' elements (shown in Chapter 10, Tables 10.1 and 10.3) represent the *minimum* that is currently considered reasonable to meet requirement L of the Building Regulations. Designers are, of course, at liberty to exceed the standard by providing a higher level of thermal insulation and there are good practical reasons for doing so, not least because upgrading insulation at a later date is both disruptive and expensive.

One of the problems with insulating over and above the currently required level is being able to prove that the work has been done to a better standard at a later date. This becomes significant when an EPC is sought for the dwelling. An EPC is mandatory if the property is to be sold or let. A higher level of insulation would improve the energy rating – and potentially the market value – of the dwelling. EPCs are intended to provide precisely this fiscal stimulus.

The most commonly used approach for measuring the energy performance of a dwelling is the reduced data SAP (Standard Assessment Procedure) (RdSAP). The difficulty with this approach is that the values for thermal elements used in the assessment are based on those prevalent at the time of construction, not what was actually built.

However, documentary evidence can be taken into account when a survey for an EPC is carried out. Householders who choose to insulate to a higher standard, therefore, should request that the contractor provides them with a record of the types and thicknesses of insulation provided at the time of construction. This might take the form of drawings, a written specification and receipts for materials purchased. This information should be retained by the householder and presented to the person carrying out a future energy assessment.

Note that the need for an energy assessment and an EPC is triggered only by a change of ownership or tenure; an EPC is not mandatory simply because a loft is converted.

## TOWARDS ZERO CARBON

High energy prices, concerns about rising emissions and questions over the UK's energy security are continuing to spur efforts to improve the efficiency of Britain's housing stock. As noted in Chapter 10, one of the ways the government is pursuing energy efficiency goals is through progressive tightening of the Building Regulations and Approved Document guidance.

Applying higher standards to new buildings presents few technical problems. Upgrading the existing housing stock is, by contrast, a far trickier task and until recently, practical examples of how this might be achieved were relatively thin on the ground.

This is beginning to change, however, with a growing number of projects up and down the country designed to highlight the potential of low-carbon technologies and improved energy efficiency. These include the Building Research Establishment's Victorian terrace at the organisation's Watford site, one of more than 300 refurbishment projects in which BRE has played a part.

Schemes of this sort not only show what can reasonably be achieved with existing buildings, but also – and equally importantly – they help to raise the profile of refurbishment and stimulate public interest. Details about individual projects are compiled by the National Refurbishment Centre, with case studies published on its refurbishment portal.

The majority of these projects are whole-building refurbishments. The reality, though, is that most improvements to property will continue to be captured on a piecemeal basis through alterations such as loft conversions. Financial constraints and purely practical considerations mean that whole-house refurbishment will remain a relative rarity for the foreseeable future, at least for the 70% of the housing stock that is owner-occupied: few householders are able to vacate a dwelling for the months or even years it can take to carry out a total refurbishment.

# The Zero Carbon Loft

As noted above, routine alterations to dwellings, such as loft conversions, will play an increasingly important part in upgrading the existing housing stock. In most cases, a loft conversion constructed to current standards is between seven and eight times more thermally efficient than the building beneath it. By incorporating additional insulation and low-carbon technologies, even greater improvements in efficiency and environmental performance can be achieved. These have the potential to meet – and exceed – current zero carbon standards.

The Zero Carbon Loft, developed by low-carbon design specialists Green Structures, is a case in point (Fig. 11.2). The dwelling chosen for the company's pilot project – a Victorian terraced house in Ealing, west London – is typical of the type of building that is routinely converted in towns and cities throughout England and Wales.

The project meets Passivhaus and better standards of insulation throughout (i.e. limiting elemental U-values equal to or less than $0.15\,\text{W/m}^2\,\text{K}$ for walls, roofs and floors). Walls contain between 150 and 175 mm polyurethane (PUR) insulation, more than twice the current Building Regulations standard. South-facing walls contain a heat-reflecting membrane to help keep the building warmer in winter and cooler in summer.

Primary roof insulation is provided by rigid PUR (225 mm), with a live sedum roof covering that contributes to overall thermal performance (Fig. 11.3). Windows are triple glazed with warm edge spacers. The entire loft area incorporates a vapour/air tightness

**Fig. 11.2** The Zero Carbon Loft in Ealing, west London. For a colour version of this figure, please see the colour plate section.
Courtesy Green Structures.

**Fig. 11.3**  The Zero Carbon Loft incorporates a green roof, while the roof of the back addition (outrigger) has been opened up to create a terrace. For a colour version of this figure, please see the colour plate section.
Courtesy Green Structures.

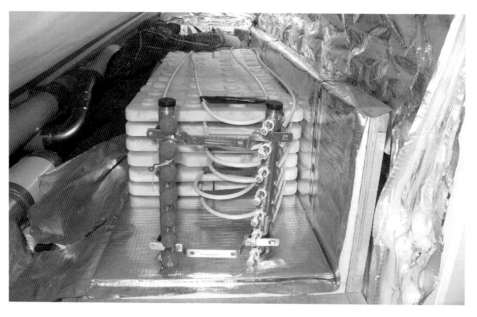

**Fig. 11.4**  Thermal accumulator stores surplus heat energy. For a colour version of this figure, please see the colour plate section.
Courtesy Green Structures.

**Fig. 11.5**   Energy-efficient LED lighting is used both inside and outside the conversion. For a colour version of this figure, please see the colour plate section.
Courtesy Green Structures.

**Fig. 11.6**   Zero Carbon Loft roof terrace during construction. For a colour version of this figure, please see the colour plate section.
Courtesy Green Structures.

barrier; all insulation joints are sealed with silicone to limit leakage and accommodate small structural movements.

The Zero Carbon Loft includes a number of significant innovations that further enhance the building's efficiency and comfort. One of these is the use of DuPont Energain panels inside the conversion. These are relatively thin (5.26 mm) and contain a phase change material (PCM), which means that heat is released or absorbed as the material within the panels – a wax compound – shifts between solid and liquid states.

When temperatures rise within the conversion, the panels act as a heat sink, helping to keep the space cool by absorbing and storing heat energy. They also provide passive heating: when temperatures fall again, stored heat energy is released back into the room. These panels are also built in to a floor lying behind a south-facing window. As well as saving energy, the panels help to create a more comfortable environment for occupants by ironing out the rapid rises and falls in temperature that are associated with highly insulated structures.

The conversion also makes use of an innovative solar heating system with externally mounted solar collectors. In conventional ST systems, the solar collectors simply heat tank water for washing and bathing; in the Zero Carbon Loft, however, this energy can be used for both water and space heating. At the heart of the system is a sophisticated thermal accumulator (Fig. 11.4) that incorporates a PCM (similar to that used in the internal wall, floor and roof panels), so surplus energy can be stored and used over a period of days if necessary. The system is fully integrated, with monitoring and control of all elements provided by a building management system.

The benefits of the approach adopted are significant, both in terms of long-term operating savings and environmental impact.

### Gas consumption

Cut by approximately 80% through improved insulation, passive solar design, passive solar heating and reduced hot water demand.

### Electricity consumption

Reduced by 60% (80% per square metre when the increased size of the building is taken into account). This has been achieved through the use of LED lighting (Fig. 11.5) with presence sensors, the replacement of inefficient appliances and smart energy consumption displays.

### Water consumption and run-off reduction

Freshwater consumption reduced by 50% through greywater recycling, low-flow shower and tap fittings, and efficient appliances. Thirty per cent reduction in rainwater run-off via sedum roofing combined with garden water butt.

# Appendix A  Specification

A building specification should state clearly (a) what materials are to be used and (b) how and where those materials are to be used. In the case of relatively small building works, such as loft conversions, the amount and type of detail in a specification (if one is written) will depend primarily on how, and by whom, the work is to be carried out.

Where a loft conversion is designed by an architect, for example, the specification will incorporate both regulatory detail (to satisfy the building control body) and a high level of technical detail (to allow contractors to tender for the job and to assist builders during construction).

However, most conversions are carried out by specialist loft companies that provide both design and building services. As work in these cases is seldom put out to tender, a scaled-down version of the specification is produced; the information provided is sometimes described as 'specification notes', 'construction notes' or a 'building control specification'. It may take the form of an independent document or, more commonly, extended annotations to drawings.

Where work is being carried out under the full plans procedure, the information contained in the specification or other notes is pivotal in demonstrating whether or not a proposal complies with Building Regulations. The specification must therefore be clear and unambiguous.

The specification of appropriately certified proprietary items will help to indicate compliance. Examples include VELUX windows with model numbers, casements with Window Energy Rating certificates, insulation with the manufacturer's product code and thickness, and fire doors with BWF CERTIFIRE certification. Where the specification relates to a system rather than a single specific product (a built-up felt roof, for example), indicating the materials to be used, the mode of application and relevant standards would be appropriate. Structural engineering calculations are usually independent of the specification.

Note that an appropriately detailed specification can be valuable from a householder's point of view if, for example, the conversion is being insulated to a higher standard than that prevalent at the time of construction. Where an Energy Performance Certificate (EPC) is sought for the property at a later date, the original specification may be of assistance in an energy assessment survey (see Chapter 11).

The following specification is based on a typical hip-to-gable loft conversion carried out by a specialist conversion company under permitted development during 2011. The conversion adds a new storey to an existing two-storey dwelling. Two habitable rooms and a bathroom with WC are provided. This specification is provided for the purposes of illustration only.

*Loft Conversions*, Second Edition. John Coutts.
© 2013 John Coutts. Published 2013 by Blackwell Publishing Ltd.

## BEAMS

- Install universal beams and universal columns in accordance with the structural engineer's design. Beam ends to bear on padstones or mild steel bearing plates. Plates or padstones to be built into brickwork and bedded in 10 mm thick mortar. Reinstate brickwork around beam ends and pack voids well with mortar.
- Floor beams to provide 30-minute fire resistance by coating with intumescent paint or by encasing in 2 no. layers of 12.5 mm plasterboard with staggered joints taped and filled. All new beams to stand 25 mm (minimum) clear of existing ceiling.

## NEW FLOOR

- Fix new $47 \times 220$ mm strength class C24 structural floor joists at 400 mm c/c. Joist ends to be fixed directly into beam webbing with solid web blocking, or fixed to beam bearers or trimmers with Simpson Strong-Tie JHA450 adjustable joist hangers fully nailed.
- Solid strutting to be provided between joists spanning in excess of 2500 mm. Joists to be 40 mm clear of chimney breasts or 200 mm clear of flue. Double joists to be provided under non-load-bearing partitions. Multiple timbers bolted at 600 mm c/c with M12 grade 4.6 bolts, washers and toothed plate connectors.
- Existing ceiling joists to be strapped to new floor joists with $20 \times 1$ mm galvanised steel band.
- 100 mm mineral wool between existing ceiling joists supported on chicken wire where existing ceiling is lath and plaster or 9.5 mm plasterboard.
- Lay 22 mm T&G flooring-grade chipboard throughout, including eaves storage space over existing ceiling joists. Floor to bathroom to be laid with moisture-resistant chipboard.

## WALLS – DORMER

- $47 \times 100$ mm vertical studs (C16) fixed to sole and head plates and header beams. External sheathing to be 12 mm OSB3 cloaked with breather membrane.
- $38 \times 25$ mm treated softwood tiling battens minimum length 1200 mm to be fixed through membrane and OSB to studwork at 114 mm gauge (maximum) with plain tile hanging ($265 \times 165$ mm). Fix soakers and flashings as required to all roof and window junctions in Code 4 lead sheet.
- Where external stud wall has an area greater than 1 m² and is within 1000 mm of a relevant boundary, fix 9 mm Promat SUPALUX boards between sheathing material and underlay/tile battens to the outside, and finish internally with 12.5 mm plasterboard with 3 mm gypsum plaster finish.
- Friction fit Celotex GA4075 rigid insulation to all stud voids flush with the back of studs, with 25 mm cavity and Celotex TB4012 to inside face of studs. Insulation joints to be taped and perimeter sealed with mastic to provide a vapour control layer. Finish with 12.5 mm plasterboard with plaster skim coat.

## WALL – HIP TO GABLE (SOLID MASONRY)

■ Existing hip and jack rafters to be removed and replaced with new 47 × 100 mm common rafters at 400 mm c/c.
■ New solid gable wall in stock bricks to match existing with 77.5 mm thermal laminate (65 mm insulation/12.5 mm wallboard) internally on dabs with skim plaster finish.
■ Re-cover new pitched portion of roof with tiles recovered from hip on underlay on battens.
■ Restraint straps (30 × 5 mm) to be fixed to gable end and to three rafters/floor joists at 2 m (maximum) centres.

## WALLS – PURLIN AND PERIMETER

■ Purlin and perimeter walls to be 47 × 100 mm vertical studs (C16) at 400 mm c/c or centres suitable to rafters with 47 × 100 mm head and base plates.
■ Friction fit Celotex GA4075 rigid insulation to all stud voids flush with the back of studs, with 25 mm cavity and Celotex TB4012 to inside face of studs. Insulation joints to be taped and perimeter sealed with mastic to provide a vapour control layer. Finish with 12.5 mm plasterboard with plaster skim coat.

## WALLS – INTERNAL PARTITIONS

■ Internal partitions to be 47 × 100 mm vertical studs at 400 mm c/c fixed to head and base plates with staggered noggings. Both sides faced with 12.5 mm Gyproc SoundBloc (minimum mass 10 kg/m$^2$) and plaster skim coat finish. Voids to be filled with snug-fitting 100 mm mineral wool to conform to guidance in Approved Document E.

## WALLS – STAIR ENCLOSURE

■ 47 × 100 mm vertical studs at 400 mm c/c fixed to head and base plates with staggered noggings. Both sides faced with 12.5 mm plasterboard with plaster skim coat finish to provide 30-minute fire resistance.

## WINDOWS AND VENTILATION

■ Roof window openings to be fully trimmed with 2 no. 47 × 100 mm C24 rafters. Trimmers also provided above and below opening.
■ All windows to achieve a minimum of Window Energy Rating Band C or better, or a U-value of 1.6 W/m$^2$ K.
■ Guarding to be provided where sill height is less than 800 mm above finished floor level unless window is in roof slope.

- All habitable rooms to have openable windows with rapid ventilation equivalent to 5% of floor area. Background ventilation of 5000 mm² (minimum).
- Bathroom to be fitted with mechanical ventilator to achieve 15 litres/second extraction connected to light switch and set to overrun for 15 minutes if window not provided.

## EXISTING AND NEW ROOF SLOPES – INSULATION

- Celotex GA4050 friction fitted between rafters with Celotex GA4070 beneath rafters. Joints of insulation beneath rafters to be taped to form vapour control layer. Fix 25 mm battens in line with rafters and fix 12.5 mm plasterboard. U-value 0.18 W/m² K.

## EXISTING AND NEW ROOF SLOPES – VENTILATION

- Existing roof structure to be provided with proprietary ventilators or vents to existing eaves to achieve equivalent continuous ventilation of 25 mm (eaves). Tile or ridge vents to give equivalent of 5 mm continuous ventilation (ridge). Provide insect mesh to all vents.

## FLAT ROOF – WARM DECK

- 12 mm mineral chippings bedded in bitumen on 3 no. layers of polyester base roofing felt (BS 747) hot bedded and laid to BS 8217. Base layer to be single-ply G3 felt with 25 mm diameter holes equally spaced and partially bonded to Celotex TD4126 warm deck insulation boards bedded on mastic beads at joints according to manufacturer's guidance to provide vapour check. U-value 0.18 W/m² K.
- Roof deck on tapered firrings to achieve 1:40 fall on 47 × 170 mm flat roof joists (C16) at 400 mm c/c. Flat roof joists to be fixed into roof beam webbing. Roof joists at dormer face to be fixed with holding-down straps to window header (2 no. 47 × 170 mm). Ceiling to be finished internally with 12.5 mm plasterboard and skim coat plaster.

## FLAT ROOF – COLD DECK

- 12 mm mineral chippings bedded in bitumen on 3 no. layers of polyester base roofing felt (BS 747) hot bedded and laid to BS 8217. Base layer to be single-ply G3 felt with 25 mm diameter holes equally spaced and partially bonded to roof deck.
- Roof deck to be 18 mm exterior plywood on tapered firrings to achieve 1:40 fall on 47 × 170 mm flat roof joists (C16) at centres to match existing rafters. Celotex XR4120 rigid insulation to be fitted between roof joists maintaining minimum 50 mm air gap to cold side. Fix Celotex TB4030 beneath joists. Ceiling to be finished internally with 12.5 mm plasterboard and skim coat plaster.
- Flat roof joists to be fixed to top of ridge beam to provide continuity of ventilation. Joists at dormer face to be fixed with holding-down straps to window header (2 no.

$47 \times 170$ mm). Fascia or soffit vents to provide equivalent of 25 mm continuous ventilation.

## FLAT ROOF – GENERAL

■ Fix softwood/PVC-U soffit and fascia boards with 112 mm gutter set to 1:350 fall to 68 mm downpipe discharging to existing rainwater drainage system.

## STAIRCASE

■ New staircase to be in accordance with BS 585-1:1989. Estimated floor-to-floor of 2756 mm gives 13 equal risers of 212 mm and equal going of 245 mm at 42° (maximum) pitch. Staircase width 750 mm. 12.5 mm plasterboard to stair soffit.
■ Stair headroom minimum 2000 mm depending on position. Handrail set at 900 mm minimum above pitch line. No part of balustrade or staircase to allow passage of a 100 mm sphere.

## ELECTRICAL

■ Provide low energy light fittings that number not less than three per four of all light fittings in the area affected by the building work.
■ Electrical installation to be designed, installed, inspected and tested in accordance with the requirements of BS 7671 (IEE Wiring Regulations) and Approved Document P *Electrical safety.*

## DRAINAGE

■ Bath, basin and shower waste pipes to be 40 mm diameter. Runs up to 4 m to be 50 mm diameter all connected separately with water seal traps to existing SVP or new 100 mm common branch pipe. Cleaning access to be provided at change of pipe direction. Opposed connections to SVP to be offset at least 200 mm.
■ Vent pipe within 3 m of any openable window to be extended 900 mm above window and provided with cage or perforated cover to conform to guidance in Approved Document H.

## FIRE

■ All doors to protected stair to provide fire resistance to FD 20 standard with BWF CERTIFIRE or equivalent certification.
■ Door glazing and fanlights within protected stair enclosure to be fitted with fire-resisting glazing.

- Provide mains-operated linked smoke alarms fitted with battery reserve supply to all floors to conform to guidance in Approved Document B.

## WATER TANKS

- Provide replacement water storage tank with valves, tank lid and pipework as necessary.
- Remove and reposition central heating feed and expansion tank.

# Appendix B The Building Regulations: appeals and determinations

The mechanisms for handling appeals and determinations described in Chapter 2 may be used when a dispute arises between a building control body and a person carrying out building work. Recourse to either of these procedures is generally considered to be a last resort and outcomes seldom favour the applicant, although there are occasional exceptions.

The number of appeals and determinations is relatively small, with just over 170 cases considered in England and Wales between 1998 and 2011. However, a third of the cases in this period were concerned with aspects of loft conversion in single-family dwellinghouses. This is a high proportion when loft conversions are considered as part of construction activity as a whole and to an extent this reflects the difficulties encountered when adapting existing buildings, or parts of them, for new purposes.

For reasons described elsewhere in this book, providing an appropriate means of escape and safe stair access is not always straightforward. It is therefore not surprising that most of the appeals and determinations concerning loft conversions deal with these requirements; three-quarters of the cases concern Requirement B1 *Means of warning and escape*. The volume of cases emphasises the sometimes intractable problems of internal layout, but arguments advanced by applicants and appellants also indicate a widespread misunderstanding of Requirement B1 and particularly the guidance that supports it.

Determination and appeal decisions should not be regarded as setting a precedent. The Secretary of State is required to consider all cases on their individual merits and some factors that are specific to previous cases are not relevant to subsequent ones.

However, the arguments advanced in both determinations and appeals proceedings provide a valuable insight into the way in which Building Regulations requirements and the supporting guidance are interpreted. Decisions and determinations by the Secretary of State thus serve to reinforce and sometimes clarify the application of existing guidance.

As explained in Chapter 2, the determination and appeals procedures have distinct roles. In a determination, a decision is made on whether or not a proposal complies with specific Building Regulations requirements. In an appeal, the appellant accepts that a proposal does not comply and seeks a relaxation or dispensation of a requirement.

The following pages contain extracts from determination and appeal letters published by the Secretary of State, all of which concern loft conversions. Where appropriate, the letters are reproduced in full.

*Loft Conversions*, Second Edition. John Coutts.
© 2013 John Coutts. Published 2013 by Blackwell Publishing Ltd.

## POTENTIALLY HABITABLE ROOMS AND ENCLOSURE

Difficulties frequently arise where a stairway discharges directly into an open-plan layout, effectively a habitable room, at ground-floor level. However, it is stressed that the relationship between the stairway and habitable rooms may become critical *elsewhere* in a dwelling as well. For example, the need to provide space for a new stair to a loft conversion means it is sometimes necessary to reconfigure rooms and landings on the floor immediately below the new storey.

In determination 45/1/194 (1 June 2001), the Secretary of State considered, amongst other things, the possible use of a large floor area ($5 \times 3$ m) at the head of a first floor stair. The applicant's client contested that this space could only function as a landing. While noting that there was no definitive way of deciding whether or not the area would be regularly used for habitable purposes, the Secretary of State noted that:

> … it is a reasonable assumption that the space is likely to be used as more than a landing and therefore, as an open-plan first floor layout, could present a considerable fire risk and threat to the occupants of the building.

A similar principle applied in determination 45/1/215 (13 August 2004). Because it was not possible to position a new stair to a loft conversion above the existing stairway, it was proposed to locate it within a first floor room previously used as a bedroom with separation from the rest of the first floor provided by a new 30-minute fire-resisting door.

The applicant indicated that this room, with approximate measurements of $4.5 \times 3.5$ m, would be used as a dressing room and would not be used as a bedroom independently of the proposed second floor accommodation. In addition, the applicant suggested that the case rested on whether there is a maximum size for an enclosure in which the stair may be located and whether the provision of wardrobes or cupboards within a stairway alters the space from an enclosure to a habitable room.

The Secretary of State acknowledged that there will often be some form of fire loading within circulation routes in domestic situations, but that the risk of a fire starting increases when people are engaged in activities other than simply travelling from one room to another. In terms of making a judgement as to whether a space should be regarded as part of a protected stairway, or as a room likely to be regularly used for habitable purposes, the Secretary of State commented that:

> … some guidance can be derived from the scale of the building; the number of rooms and the usability of the space; and the number and position of the doors which open off the area. In this case the dressing room is similar in size to the adjacent bedrooms at first floor level and, as a dressing room, would clearly be used for purposes other than circulation.

## DEFINING A HABITABLE ROOM

For the purposes of Approved Document B, a habitable room is described as one which is:

> … used, or intended to be used, for dwellinghouse purposes …

A storage area would not normally be considered to be a habitable room. But simply describing a planned loft conversion as 'storage space' would not remove the obligation to provide an appropriate means of escape (a staircase rather than a ladder) where it is likely that the storage space in question could be used for habitable purposes by current or future occupants.

In a determination letter dated 23 August 2006 (45/1/224), which concerned a proposed loft conversion (designated as a storage area) in a two-storey dwelling, the Secretary of State noted that:

> … it is a reasonable assumption that the roof space in this case is likely to be used for more than storage, if not by your client then by future occupants of the building.

Points that might be taken into account when considering whether the roof space is a habitable room could include:

> … its size (particularly in relation to the rest of the building), whether it has electrical services (e.g. power sockets etc.), is plastered, has a stair (of any type), and, possibly, if there is a window even if the intention at the time is only for it to be used for storage.

The proposed conversion – which the Secretary of State determined did not make appropriate provision for means of escape in case of fire – included a boarded-out floor, electrical services, plasterboard finished walls and three windows, including a dormer window.

## DEFINITION OF A STAIR

It is sometimes difficult to achieve adequate headroom in loft conversions. While there is no guidance on minimum floor to ceiling height within rooms, Approved Document K *Protection from falling, collision and impact* sets out minimum clearances for stairs and landings. While these are unequivocal, defining exactly what constitutes a 'stair' for the purposes of the guidance is less clear, particularly in conversions where there are relatively complex configurations of steps and landings.

This was the subject of two appeals against refusal to relax requirement K1, both dated 6 February 2004 (references 45/3/165 and 45/3/161). The text of appeal 45/3/165 is reproduced below.

### The appeal

This appeal relates to completed building work to create a 31.5 m² bedroom with integral en suite shower and WC in the roof space of a four-bedroom (previously two-storey) detached house approximately 8 m × 6 m in area (i.e. a loft conversion). The roof is of pitched, single ridge construction running between the flank walls; and the new second floor room has been created by breaking open approximately three-quarters of the length of the rear pitch from eaves level and constructing a dormer framework containing three separate windows […].

Access to the new room is by a timber stair installed over the ground to first floor stair. At the foot of the new stair are two winders and at the top there is a quarter 'drop landing' giving access via an additional step, facing the stair, to the new room. The room is protected by an inward opening fire door located on the top of the additional step.

These proposals were the subject of a full plans application which was conditionally approved. This included a condition to ensure a 2 m minimum headroom at the head of the stair. However, it is understood that the floor and dormer roof were not built to the levels indicated on the approved plan with the result that the Borough Council considers that the headroom for the stair at the top (i.e. additional) step is not in compliance with Requirement K1.

However, you took the view that Requirement K1 should not be applied to the question of headroom from this additional step up from the 'drop landing' into the new room. You therefore applied to the Borough Council for a relaxation of Requirement K1 which was refused. The Council then issued a notice of contravention in respect of Requirement K1 requiring corrective works within 28 days. It is against the refusal to relax Requirement K1 that you appealed to the Secretary of State.

## The appellant's case

You consider that the stair to the new second floor room comprises the main flight and the drop landing, and that you have provided approximately 2 m headroom throughout. You argue that the headroom from the additional step at the door into the new room should not be taken into consideration because it is not part of the flight.

You also state that you have discussed the form of construction with an adjacent Borough Council's building control division who consider that your proposals comply with the current Building Regulations.

## The Borough Council's case

The Borough Council has considered the definition of a 'stair' given in Approved Document K *Protection from falling, collision and impact*, which is: 'a succession of steps and landings that makes it possible to pass on foot to other levels'. The Council regards the landing, the additional step at the doorway to the new second floor room, and the doorway itself as forming part of the stair. The Council points out that the headroom under the doorway is approximately 1.76 m and is approximately 1.86 m under the ridge in the new room. Given that Approved Document K recommends that 2 m is adequate headroom, the Council does not consider that Requirement K1 has been complied with.

The Borough Council also considers that the door sweeping across what it regards as the upper landing contravenes the guidance in paragraph 1.16 of Approved Document K, which says that: 'To afford safe passage landings should be clear of permanent obstruction.'

To support its case the Borough Council has enclosed a copy of an appeal decision by the Secretary of State in a case involving headroom which he issued in 1996.

## The Secretary of State's consideration

Falls on stairs in dwellings are a very common type of accident resulting in about 500 deaths per year and many thousands of injuries. The Secretary of State therefore considers that good stair design makes an essential contribution to life safety.

In considering this appeal the Secretary of State has first considered to what degree the proposed stair may fall short of compliance with Requirement K1, thereby potentially warranting a relaxation of this requirement.

Requirement K1 says that: 'Stairs, ladders and ramps shall be so designed, constructed and installed as to be safe for people moving between different levels in or about the building.' The guidance in Approved Document K gives solutions for common situations, but loft conversions often present particular problems which have to be considered on their individual merits. The overriding consideration is the safety of the stair user.

For the purposes of Part K *Protection from falling, collision and impact*, a stair is defined as: 'A succession of steps and landings that makes it possible to pass on foot to other levels.' You and the Borough Council disagree as to whether the stair ends at the foot of the additional step and new door, or continues into the proposed new second floor room. In the Secretary of State's view, if there was no door, then the stair should be considered to include the single additional step together with the landing beyond formed by the adjacent part of the floor of the new room. However, in this case there is a door, and the Secretary of State considers that it is reasonable to regard the stair as ending at that door, even though this landing is not strictly at the upper level. This is because the door provides a clear barrier to progress, which requires the user to stop and open it before proceeding. The door also provides a clear marker for the change in level at the single additional step.

The guidance on performance on page 5 of Approved Document K makes it clear that Requirement K1 will only be applicable to differences in level of more than 600 mm. Given that it is the Secretary of State's view that the stair ends at the additional step and the door, it follows that the step is not required to comply with Requirement K1 of the Building Regulations because the difference in levels here is less than 600 mm. It also follows that because the floor area of this part of the new room cannot be considered to be a landing, the room height of this part of the room is not subject to the Building Regulations.

The Secretary of State has also noted the Borough Council's reference to a previous appeal decision made in 1996, which the Council contends supports its case. However, the Secretary of State is required to consider all cases on their individual merits. He considers that loft conversion cases can pose different questions, and issues specific to previous cases will not necessarily be relevant to subsequent ones.

In a more general context than the precise application of the Building Regulations, the Secretary of State recognises that the overall concern of the Borough Council is the safety of those using the stair. Although there is adequate headroom to the stair itself, the Council's concern is with the limited headroom of 1.76 m under the doorway and approximately 1.86 m under the ridge in the new room. These figures are in contrast to the figure of 2 m which is defined as adequate headroom on the access between levels in paragraph 1.10 of Approved Document K. However, paragraph 1.10 recognises the constraints which may exist in loft conversions and says: 'For loft conversions where there is not enough space to achieve this height, the headroom will be satisfactory if the height measured at the centre of the stair width is 1.9 m reducing to 1.8 m at the side…'

Although the Borough Council's concerns over the limited headroom relate to construction elements which in the Secretary of State's view do not fall to be controlled under the Building Regulations (i.e. room height), he does consider it appropriate to comment as follows. As a general principle he takes the view that where it is necessary to use a drop landing, an additional step – or additional step and door – should preferably be at right angles to the direction of the main stair flight, and that the length of the drop landing should be of sufficient length to enable the 90° turn to ascend the step to be made some distance away from the top of the main flight. Such a design principle should minimise any hazard and risk of falling. When ascending the additional step any person who did bump their head is likely to do so in a position of 90° to the flight, thus increasing the chances of regaining their balance on the drop landing as opposed to falling back down the stair flight.

Notwithstanding the above general comments regarding optimum design for life safety, the Secretary of State does accept that loft conversions can present constraints on stair design, particularly in terms of headroom. As noted above, this is acknowledged in Approved Document K. Thus although the Secretary of State considers that there may be potential to further improve the design safety of this particular stair, he considers that the risk of harm to the users of the stair is acceptably small, given that they will generally be familiar with the layout. He therefore takes the view that the stair as constructed offers a reasonable level of safety and therefore complies with Requirement K1. It follows that he considers it would be neither appropriate nor necessary to relax Requirement K1 in order to secure the compliance of the existing stair.

### The Secretary of State's decision

The Secretary of State has given careful consideration to the facts of this case and the arguments put forward by both parties [...]

However, you have appealed to the Secretary of State in respect of the refusal by the Borough Council to relax Requirement K1. The Secretary of State considers that compliance with Requirement K1 makes an essential contribution to life safety and as such he would not normally consider it appropriate to relax it, except in exceptional circumstances. Moreover, because in the particular circumstances of this case he considers that your building work complies with Requirement K1, there would appear to be no prima facie case to relax the requirement in any event. Therefore, taking these factors into account, the Secretary of State has concluded that it would not be appropriate to relax Requirement K1 *Stairs, ladders and ramps* of Schedule 1 to the Building Regulations 2000 (as amended). Accordingly he dismisses your appeal.

## STAIR DESIGN – PITCH AND GOING

Stair designs that diverge from established patterns are potentially problematic and this is borne out in the following appeal. In an appeal against refusal to relax Requirement K1 dated 4 February 1999 (reference 45/3/130), a Borough Council had served a notice under section 36 of the Building Act 1984 to remove or alter a staircase in order for it to comply with the Building Regulations.

The appellant's staircase comprised four winders at both top and bottom with an intermediate straight section. The Borough Council considered the stair as a unit to be unsatisfactory, because, in its view, the pitch exceeded 42°, the going varied from step to step and the going of some of the steps was considered to be inadequate.

The appeal was dismissed on the grounds that Requirement K1 can be a matter of life safety. However, the Secretary of State took the view that this particular stairway would comply with the requirement if modified:

> The appropriate design and measurements for winder stairs are covered in BS 585: 'Wood stairs' Part 1: 1989. This recommends that the going of the straight part of the flight should be as recommended in BS 5395 ('Stairs, ladders and walkways' Part 1: 1977 (confirmed November 1984) 'Code of practice for the design of straight stairs'), and that the centre going of the winders should be uniform, and not less than the going of the straight part of the flight. The smallest going recommended in BS 5395: Part 1 is 225 mm. However, Part 2 of that BS deals with spiral and helical stairs and allows centre goings as small as 145 mm in situations such as this.
>
> Clearly, your stair falls well short of the guidance in BS 585, but it is within the limits for spiral stairs. Small spiral stairs are considered to be safer than straight or winder stairs because, in the event of a fall, the user is likely to fall towards the guarding/handrail, thus providing them with an opportunity to regain balance.

Although not shown in the photographs you have submitted, you state that additional features have been added to the stair including extra handrails, grab rails and handles. This would improve the safety, especially if the handrails are continuous.

In loft conversions such as this, the guidance in Approved Document K takes account of the limited space available and the light use – mainly by people who are familiar with the stair – and suggests provision of either a fixed ladder or an alternating-tread stair. In the Department's view the stair as proposed and installed would be as safe as these alternatives and in the context of this particular situation complies with Requirement K1.

## STAIR DESIGN – SELECTIVE APPLICATION OF REGULATIONS

The risks posed by stairs with open risers and without guarding is considered in an appeal against refusal to dispense with Requirement K1 (reference 45/3/141, dated 13 January 2000). The appeal concerned stair access for a bungalow loft conversion. The stair had been designed for aesthetic effect with open risers of approximately 200 mm. A handrail was provided on one side but the stair was open on the other. The appellant was prepared to change the stair to comply with the requirements of the Building Regulations were the house to be sold.

The District Council expressed the view that the requirements of the Building Regulations

cannot be taken in part for the particular occupants at the time to choose the elements which are felt to apply to them.

This view was supported by the Secretary of State:

The Department notes and endorses the point made by the District Council that the Building Regulations cannot be applied selectively but must relate to all persons, including young children, in or about a building; and that they must be applied at the time of the building work. In the Department's view, compliance could be readily achieved in practical terms in respect of both the guarding and the open risers. Moreover, the effect of the open risers need not necessarily be completely compromised if an adequate guarding were to be installed to ensure that a 100 mm diameter sphere could not pass through the risers, as recommended in paragraph 1.9 of Approved Document K.

## VENTILATION

The increasing emphasis on creating airtight buildings means that air movements between the inside and outside of dwellings must now be systematically accounted for and managed. As a consequence, ventilation – including the provision of mechanical ventilation – is now closely checked by building control bodies when building work is carried out in both new and existing dwellings.

Arguments for *not* providing mechanical ventilation were advanced in an appeal against a refusal to dispense with Requirement F1 *Means of ventilation* (reference 45/3/197, dated 12 March 2009). However, these were refuted by the Secretary of State, who concluded that it would not be appropriate to dispense with the requirement. Although the following appeal makes reference to an earlier version of Approved Document F, the principle remains unchanged.

## The building work and appeal

The papers submitted indicate that the building work to which this appeal relates is complete and comprised a first floor rear extension and enlargement of an existing second floor loft room of a five bedroom detached property, for which planning permission was received from the Council in 2007. The work included the installation of a new en-suite bathroom on the first floor and a new shower room in the loft room (also referred to below as 'bathrooms').

A building regulations full plans application was deposited with the Council for the building work which was rejected on matters unrelated to your dispensation application and appeal. Work started on site on 26 June 2007 and a final inspection was carried out on 14 March 2008 where a number of outstanding matters were noted by the Council. Included in these items was the non-provision of mechanical ventilation in the two new bathrooms, which the Council stated was required under Requirement F1 (Means of ventilation) of the Building Regulations.

Due to the lack of mechanical ventilators in the bathrooms you were informed that a completion certificate could not be issued. As you considered that these were unnecessary you applied for a dispensation of Requirement F1 on 1 July 2008, which the Council refused. It is against this refusal that you have appealed to the Secretary of State.

## The appellant's case

You have enclosed a copy of a statement from your builders which in your and their view is 'proof that the provisions provided are adequate to secure reasonable standards of health and safety for persons in the building'. You add that you have used the new bathrooms since November 2007 and do not see any practical need for installing mechanical ventilators, which you consider will waste energy, create noise and bother your neighbours.

Your builders' statement indicates that both bathrooms are of substantial size (en-suite bathroom 37.5 cubic metres, shower room 22.5 cubic metres) and are provided with large windows within the extension. The statement adds that the following measures were provided in order to prevent the occurrence of future moisture problems:

- The size of the opening windows in the shower room is greater than 1/6 of the floor area and greater than 1/17 in the en-suite bathroom. (Appendix B of Approved Document F requires the window area to be at least 1/20 of the floor area).
- The en-suite bathroom has a frameless glass door which leaves a surrounding gap of 10 mm.
- All surrounding walls as well as floors and ceilings are fully insulated according to building regulations requirements.
- The bathroom walls are fully tiled from floor to ceiling to prevent build up of mould. Floors are also tiled and ceilings painted with waterproof bathroom paint.

In response to the Council's representations to the Secretary of State, you added that your understanding is that the issues relating to the rejection of your plans and the completed work were resolved apart from the question of ventilation in the new bathrooms. In your view the Council has not explained their position or responded to all your arguments indicated above. You added that:

- There has been little to no steam building-up since the bathrooms have been in use. When the windows are opened the steam disappears after a couple of minutes. Even if the windows were not opened, the steam would not be of any harm to health and safety as the tiling of the rooms prevents the creation of mould.

- In your view, an electrical operated ventilated system will consume electricity and the existence of a ventilation hole is a source of waste of heating energy to the outside. The Council's requirement is counter-productive to a 'green concept'.
- You believe the Council has been 'lenient' in the past in enforcing Requirement F1.

## The Council's case

As indicated above, the Council has refused your dispensation application relating to the "requirement for mechanical extractors to the two new bathrooms" as they consider that this is required by Requirement F1 of the Building Regulations and was specified in "the Architect's construction notes".

With regard to Requirement F1, the Council states that for ventilation rates for bathrooms to be effective, they should have a rapid extract capacity suggested at 15 l/s (intermittent) or 8 l/s continuous. The Council notes the case you have submitted to obviate the need for rapid ventilation but does not accept this as the most effective ventilation associated with the high vapour and relative humidity associated with bathroom usage.

## The Secretary of State's consideration

The Secretary of State has given careful consideration to the particular circumstances of this case and the arguments presented by both parties and has referred to the guidance in Approved Document F (Ventilation – 2006 edition) to assist her in reaching a decision.

The Secretary of State observes that the Building Regulations now encourage builders to make the building envelope airtight in order to save energy. This makes it more important that an adequate ventilation system is installed and used correctly in modern buildings, including in new extensions.

Approved Document F describes [...] a number of ventilation systems that will meet Requirement F1 of the Building Regulations when adding a room where moisture is produced – such as a bathroom or kitchen – to an existing building. The Approved Document advocates providing all three of the following types of ventilation to a bathroom or shower room:

- extract ventilation (for example an intermittent extract fan) to remove water vapour before it can condense on cold surfaces and possibly support mould growth, and also to minimise the spread of water vapour to the rest of the building;
- background ventilation (for example a window trickle vent) to provide a small amount of fresh air continuously;
- purge ventilation (for example an openable window) to remove high concentrations of pollutants and water vapour arising from occasional activities such as painting and decorating, and to improve comfort in hot weather.

The Council has referred to 'the Architect's construction notes' in this case which specified the use of mechanical extractors, i.e. intermittent extract fans, to provide the extract ventilation in the new bathrooms. You took the view that the fans were not needed but, as indicated above, their purpose is to minimise condensation and the spread of water vapour throughout the building which is especially important in cold weather when windows are closed.

The Secretary of State appreciates that the walls and floors of the bathrooms have been extensively tiled and that the windows are larger than required for the purposes of purge ventilation. However, she does not consider this to be sufficient reason to dispense with Requirement F1 of the Building Regulations, and for the reasons given above feels it is important to provide some form of extract ventilation in the bathrooms.

The extract ventilation may be provided either by an intermittent extract fan or, if fan noise and energy consumption are a concern, by a passive stack ventilator [...] which does not require a fan or use any electricity.

### The Secretary of State's decision

As indicated above, the Secretary of State considers that compliance with Requirement F1 is important for the reasons explained and she considers that a sufficient case has not been made to dispense with the requirement in this case. She has therefore concluded that it would not be appropriate to dispense with Requirement F1 (Means of ventilation) in Part F (Ventilation) of Schedule 1 to the Building Regulations 2000 (as amended), in relation to the need for adequate means of ventilation in the two new bathrooms in question. Accordingly, she dismisses your appeal.

# Appendix C  Planning and curtilage

There is no statutory definition of curtilage, and local planning authorities interpret its meaning differently. Some councils now require planning permission for raising a party wall at full thickness on the grounds that half the wall lies *outside* the curtilage of the building being extended. Others consider that the curtilage includes the whole thickness of the party wall and issue Lawful Development Certificates confirming that such development is permitted.

Reasons for *excluding* the raising of a party wall from permitted development were advanced in a 1997 appeal concerning a full-width conversion in a mid-terrace house in Bath (reference APP/C/96/F0114/642257). An apparent consequence of this is that an increasing number of local authorities now require an application for planning permission for such work, and a Lawful Development Certificate (which confirms that a proposal does not require express planning permission) is refused.

However, arguments for *including* a party wall within the curtilage of a dwellinghouse in the context of a roof extension are advanced in two appeal decisions:

- APP/Q5300/X/01/1062324 (dated 11 October 2001)
  Property in the London Borough of Enfield
- APP/N5090/X/09/2108111 (dated 13 January 2010)
  Property in the London Borough of Barnet

Appeal Decision APP/N5090/X/09/2108111 is reproduced in full below. It should be noted that a planning inspector's decision is definitive only in so far as it applies to a particular property at a specific point in time – it is not case law.

### *Appeal Ref: APP/N5090/X/09/2108111*
[---] Lane, Hampstead, London NW3 [---]

- The appeal is made under section 195 of the Town and Country Planning Act 1990, as amended by the Planning and Compensation Act 1991, against a refusal to grant a certificate of lawful use or development (LDC).
- The appeal is made by Mr [-----] against the decision of the Council of the London Borough of Barnet.
- The application (Ref:- F/01371/09), dated 27 April 2009, was refused by notice dated 9 June 2009.
- The application was made under section 192(1)(b) of the Town and Country Planning Act 1990 as amended.

*Loft Conversions*, Second Edition. John Coutts.
© 2013 John Coutts. Published 2013 by Blackwell Publishing Ltd.

■ The development for which a certificate of lawful use or development is sought is loft conversion with dormer window to rear addition roof, involving raising party wall parapet (extensions to roof including dormer window to rear projection to facilitate a loft conversion).

## Decision

(1) I allow the appeal and I attach to this decision a certificate of lawful use or development describing the proposed operation, which I consider to be lawful.

## Main Issue

(2) I consider that the main issue in this appeal is whether the refusal of a lawful development certificate (LDC) for the carrying out of a loft conversion, with a dormer window to a rear addition roof involving raising a party wall parapet at [---] Lane, London NW3, was well founded.

## Reasons

(3) This application to the local planning authority for a lawful development certificate, for the installation of a loft conversion, with a dormer window to a rear addition roof involving raising party wall parapet at [---] Lane, London NW3, was made after 1 October 2008. Therefore, the provisions of the Schedule to the Town & Country Planning (General Permitted Development) (Amendment) (No2) (England) Order 2008 apply to the circumstances of this case. Class B of the Schedule, which replaced the former Part 1 of Schedule 2 to the 1995 Order, describes the enlargement of a dwellinghouse consisting of an addition or alteration to its roof as permitted development i.e. development granted planning permission by virtue of the provisions of Article 3(1) of the Town & Country Planning (General Permitted Development) Order 1995 (GPDO).

(4) Article 3(2) states that any permission granted by paragraph (1) is subject to any relevant exception, limitation or condition specified in Schedule 2. The limitations and conditions imposed upon Class B in the Schedule to the 2008 Amendment Order are extensive. The limitations in paragraph B.1 include restrictions to the height of the roof alterations not exceeding the highest part of the roof, the alterations not extending the plane of the roof slope incorporated in the dwelling's principal elevation and the cubic content of the alterations to the roof exceeding the cubic content of the original roof-space by 40 m³ in this mid-terrace house, which is not in Article 1(5) land. All of these limitations are satisfied.

(5) The conditions in paragraph B.2 are similarly met. The proposed materials are similar to those used in the construction of the original house and the edge of the roof alteration closest to the original eaves is shown to be more than 20 cm from the eaves of the original roof on the submitted drawings, which similarly indicate the windows in the side elevation to be obscure glazed and fixed shut. The local planning authority accepts that all of these conditions and limitations are satisfied. However, it points out that the Schedule to the 2008 Amendment Order is headed, "*Development within the Curtilage of a Dwellinghouse*". It contends that to build up the party wall with [the adjoining property], by raising the combined thickness of the joint wall to the two properties, would constitute development not falling within the curtilage of [-----] Lane, would therefore fail to comply with the heading to the Schedule and would thereby cause the proposed roof alterations in their entirety to require planning permission.

(6) When the application for the LDC was made to the Council, an appeal decision (Reference APP/Q5300/X/01/106324), dated 17 September 2001 and made by one of my colleagues, was submitted. This related to an application made under section 192(1)(b) of the amended 1990

Act for a loft conversion at [---] Road in the adjoining London Borough of Enfield, where, as in this instance, the only reason given for the refusal of a certificate was based on the local planning authority's belief that the extension to the party wall would not fall within Class B of what was then Part 1 of Schedule 2 to the Town & Country Planning (General Permitted Development) Order 1995.

(7)   The Inspector quoted extensively from several authorities (*Methuen-Campbell v Walters* [1979] 1 QB 525, *Dyer v Dorset CC* [1988] 3 WLR 213, *Attorney-Gen ex rel Suttcliffe & others v Calderdale MBC* [1983] JPL and *McAlpine v Secretary of State for the Environment* [1995] JPL B43). In doing so, he concluded that, on the strength of the final case cited in particular, a curtilage comprised three defining characteristics. Firstly, it occupied a small area around a building, secondly it was intimately associated with that building and thirdly it had to be regarded as part of one enclosure with the house. Where party walls are concerned, the Inspector reached the conclusion that two adjoining curtilages can overlap each other, where a party wall shared by two contiguous properties could result in the partial collapse of both if the wall were removed. He could see no reason why, with a party wall such an integral part of two dwellinghouses, their two curtilages could not overlap, because such small areas were involved, and neither can I.

(8)   I appreciate that Inspectors cannot make decisions that are binding on their colleagues or that equate in any way to judicial authority. However, I note that the Inspector in the Enfield decision was a practising solicitor, whereas I am not legally qualified in any way. In *North Wiltshire DC v Secretary of State for the Environment & Clover* [1992] JPL 955 the Court of Appeal held that an Inspector was free upon consideration to disagree with the judgement of another but before doing so he ought to have regard to the importance of consistency and to give his reasons for departure from a previous decision. I can find no reason to contradict my predecessor in this area of planning law.

(9)   The only significant change in the law in the intervening period between these two decisions was the replacement of Part 1 of Schedule 2 to the Town & Country Planning (General Permitted Development) Order 1995 by the Schedule to the Town & Country Planning (General Permitted Development) (Amendment) (No2) (England) Order 2008. However, both the original Part 1 of Schedule 2 and its 2008 replacement are headed *"Development within the Curtilage of a Dwellinghouse"*, the only matter at issue between the parties in both instances. For the reasons set out above, I concur with my colleague in the Enfield LDC case. I find that raising of the party wall between [---] and [---] Lane as a joint building exercise would be part of the development permitted by Part 1 of Schedule 2 to the Town & Country Planning (General Permitted Development) Order 1995 as amended, within the curtilage of the dwellinghouse at [---] Lane, London NW3 and a lawful development certificate should have been issued by the local planning authority to that effect.

## Conclusions

(10)   For the reasons given above, I conclude, on the evidence now available, that the Council's refusal to grant a certificate of lawful use or development, in respect of a loft conversion with a dormer window to a rear addition roof, involving raising a party wall parapet (extensions to roof including dormer window to rear projection to facilitate a loft conversion) at [---] Lane, Hampstead, London NW3 [---], was not well founded and that the appeal should succeed. I shall exercise the powers transferred to me under section 195(2) of the 1990 Act as amended and I issue a lawful development certificate accordingly.

Ian Currie
Inspector

# Glossary

**AAC**   Autoclaved aerated concrete (block).

**AD**   Approved Document (England and Wales).

**ALS/CLS**   Surfaced timber with rounded arrises widely used for studwork in timber frame construction. Common sizes 38 × 89 mm and 38 × 140 mm. (American Lumber Standard, Canadian Lumber Standards).

**BRE**   Building Research Establishment.

**BUR**   Built-up roofing. Widely used flat-roof weathering system, with membranes manufactured using polyester reinforcement and modified bitumen.

**Back addition**   Rearward projection in the nineteenth and early twentieth-century terraced dwellings. The back addition may have the same number of storeys as the main building but is subordinate in width. Ceiling heights in the back addition are generally lower than those in the main dwelling. A back addition is generally shared between a pair of houses and is thus divided along its length by a party wall; the roof often comprises a pair of independent lean-to slopes (see Fig. G.1). The back addition roof space is sometimes incorporated as part of a larger conversion. Also described as an outrigger.

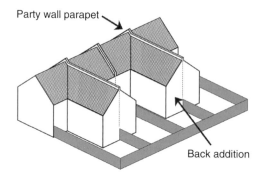

**Fig. G.1**   Back addition (outrigger).

**Bedded verge**   Roof edge finishing detail for gable ends. An oversailing rigid undercloak is formed before battens and roofing material are fixed. The void between the undercloak and the roofing material is then filled with mortar and the bedding struck off cleanly.

**Breather membrane**   Allows water vapour to escape from the building envelope but prevents the passage of liquid water. Approved Document C indicates that breather membrane, rather than traditional impermeable underlay, should be used behind vertical tile hanging.

*Loft Conversions*, Second Edition. John Coutts.
© 2013 John Coutts. Published 2013 by Blackwell Publishing Ltd.

**Butterfly roof**   This form is typically encountered in terraced dwellings of the late eighteenth and nineteenth centuries (see Fig. G.2). It has a distinctive V-shaped profile with a central gutter running at right angles to the front of the building. For the sake of uniformity in terraces, the twin slopes of the butterfly roof are generally hidden from view behind a parapet wall at the front of the building and are usually only visible from the rear. In effect, the roof comprises a pair of independent lean-to slopes. The compartment walls separating such dwellings generally project above the roof covering to form party wall parapets. The roof slopes themselves are frequently pitched at a very shallow angle: slopes of less than 20° are common. The two roof voids thus formed are generally too small to justify conversion in their own right. Also sometimes called a London roof.

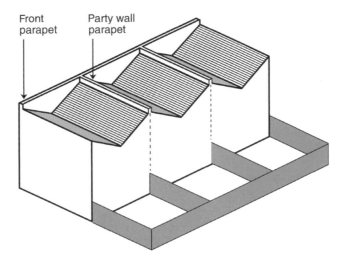

**Fig. G.2**   Butterfly roof.

**CDM**   The Construction (Design and Management) Regulations 2007.

**Certificate of Lawfulness**   (see *Lawful Development Certificate*).

**Cloaked verge**   Verge formed at gable ends using purpose-made L-section tiles. Dry (i.e. no mortar) verging systems are also available for use with slate roofs.

**Coffin tank**   The coffin tank, sometimes called a loft tank or low-level tank, is widely used in conversions where a conventional water tank cannot be accommodated (see Fig. G.3). As its name suggests, it is a long and relatively shallow tank that is suitable for fixing in confined spaces such as the eaves or the apex of the roof.

**Communities and Local Government, Department for (CLG)**   The government department with responsibility for Building Regulations and Approved Documents in England and Wales. Formerly ODPM.

**Dead load**   Load due to the weight of walls, permanent partitions, floors, roofs and finishes including services and all other permanent construction.

**Dormer**   'Dormer' describes a projecting vertical window, with its own roof, that is set into a principal roof slope. The word first entered the English language in the late

**Fig. G.3**  Coffin tank.
Image courtesy of Polytank Group Ltd.

**Coffin tank typical volumes and dimensions**

|  | 114 litres (25 gal) | 227 litres (50 gal) | 318 litres (70 gal) | 454 litres (100 gal) |
|---|---|---|---|---|
| Length (mm) | 1390 | 1650 | 1650 | 1680 |
| Width (mm) | 500 | 460 | 610 | 695 |
| Height (mm) | 310 | 475 | 500 | 580 |

sixteenth century, although the built form itself predates this. Historically, dormer windows were subordinate structures intended to provide light and ventilation to the roof space in order to make it habitable. In contemporary usage, however, the word dormer has become associated with more or less any habitable projection from a roof, regardless of its scale.

**ELV**   Extra low voltage; used with reference to lighting systems that depend on a 12 V supply via one or more transformers.

**EPBD**   Energy Performance of Buildings Directive.

**EPDM**   Synthetic sheet rubber membrane used as a roofing material.

**Energy Performance Certificate (EPC)**   An EPC provides information on a dwelling's energy use and carbon dioxide emissions. Efficiency – both current and potential – is expressed on a scale of A (very efficient) to G (not energy efficient). An EPC energy assessment is *not* currently required when a loft conversion is carried out. However, it is mandatory if the property is bought, sold or rented out. The assessment must be carried out by a member of an accreditation scheme that is approved by the Secretary of State. The price of an EPC is set by the market and depends on the size and location of the building. The certificate is valid for 10 years.

**FFL**   Finished floor level (generally on drawings).

**Fenestration**   The arrangement of windows in a building.

**Firring**   Tapered timbers that are nailed along the tops of flat-roof joists (or sometimes at right angles to them) in order to create a slope. Generally, firring is cut to achieve a fall that ranges between 1:40 and 1:80.

**Fixed building services**   Any part of, or any controls associated with: (a) fixed internal or external lighting systems, but not emergency escape lighting or specialist process lighting; or (b) fixed systems for heating, hot water, air conditioning or mechanical ventilation.

**Flaunching**   Slope on the top surface of a chimney designed to throw water.

**Floor-to-floor**   The measurement between the surfaces of different floors (e.g. an existing upper floor and the new floor in a loft conversion). This measurement is critical when commissioning a staircase for a loft conversion.

**Friction fit**   A push fit achieved by slight over-sizing of the material to be fitted (e.g. rigid insulation boards between rafters or studs).

**GPDO**   The Town and Country Planning (General Permitted Development) Order 1995. Also used to describe The Town and Country Planning (General Permitted Development) (Amendment) (No. 2) (England) Order 2008. Planning legislation. Most loft conversions are carried out under permitted development rights set out in the GPDO.

**GRP**   Glass reinforced plastic (fibreglass). Sometimes used as weathering for small dormers. Often pigmented to resemble lead or zinc. A number of specialist suppliers produce entire dormer assemblies in GRP (see Fig. G.4).

**Gable parapet**   Projection of a gable wall above the roof surface to the flank side of a building (see also *Party wall parapet*).

**Gallows bracket**   A welded mild steel bracket formed from rolled steel angles (see Fig. G.5). These are sometimes used to support masonry where an internal chimney breast has been removed. Usually secured to walls with expansion bolts. Gallows brackets are not universally accepted by building control bodies and are generally not used if the wall supporting the bracket is part of a flue.

**Glulam**   Glued laminated timber.

**HSFG**   High-strength friction grip (nut and bolt assembly).

**Habitable room (fire safety)**   For the purposes of Approved Document B *Fire safety*, a habitable room is one that is used, or intended to be used, for dwellinghouse purposes (including a kitchen, but not a bathroom).

**Habitable room (ventilation)**   For the purposes of Approved Document F *Ventilation*, a habitable room is one that is used for dwelling purposes, but which is not solely a kitchen, utility room, bathroom, cellar or sanitary accommodation.

**Hip iron**   A scroll-ended strap screwed to the base of the hip rafter to provide support for hip tiles. Sometimes called a hip hook. Generally made from 6 mm galvanised steel. Earlier examples are wrought iron.

**House longhorn beetle** (*Hylotrupes bajulus*)   Capable of causing severe damage to softwoods, particularly in roof structures. The beetle is black or dull brown in colour and between 10 and 20 mm in length. Emerges from infected timber between July and October. Larvae are up to 30 mm in length. Emergence holes are oval in shape and about 6–10 mm. Guidance in Approved Document A *Structure* indicates that softwood timber for roof construction, including ceiling joists, should be adequately treated to prevent infestation. This would also include, for example, vertical studs in a dormer wall. In general, the guidance is satisfied by the use of tanalised timber.

**Fig. G.4** GRP dormer.
Image courtesy of Euroform Products Ltd.

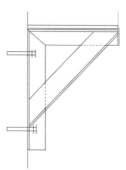

**Fig. G.5** Gallows bracket.

*Areas at risk from house longhorn beetle*
- Borough of Bracknell Forest (parishes of Sandhurst and Crowthorne)
- Borough of Elmbridge
- District of Hart (parishes of Hawley and Yateley)
- District of Runnymede

- Borough of Spelthorne
- Borough of Surrey Heath
- Borough of Rushmoor (area of the former district of Farnborough)
- Borough of Woking

**Imposed load**   The load assumed to be produced by occupancy or use, including the weight of moveable partitions and snow. Excludes wind loads. Also described as live load.

**Interstitial condensation**   Deposition of liquid water, from vapour, if it occurs within or between elements in the building (e.g. water that condenses within insulation).

**Juliet balcony**   A Juliet balcony, or 'blind' balcony, incorporates inward-opening French windows with external guarding provided by a balustrade. It is not a true balcony because there is no external floor projection.

**Lawful Development Certificate (LDC)**   Generally used to establish the lawfulness of building work carried out under permitted development legislation. Sometimes called a Certificate of Lawfulness.

**LED**   Light-emitting diode. LEDs are extremely energy efficient and for this reason they are increasingly used in domestic lighting.

**Lamp**   A light source (bulb). A lamp may be fluorescent, incandescent or LED.

**Live load**   (see *imposed load*)

**LOLER**   Lifting Operations and Lifting Equipment Regulations (1998). Legal requirements relating to the use of lifting equipment.

**London roof**   (see *butterfly roof*).

**Low-E glass**   Low-emissivity glass allows heat and light to pass into a building but limits heat loss from the building. This is achieved by applying a thin metallic oxide coating to one side of the glass during manufacture. There are two principal types of low-E glass coatings: hard (emissivities in the range from 0.15 to 0.2) and soft (emissivities in the range from 0.05 to 0.1). By comparison, the emissivity of plain glass is about 0.89.

**MoE**   Means of escape (from fire). Usually refers to an emergency egress window.

**MS**   Mild steel (on drawings).

**Material alteration (Building Regulations)**   The meaning of material alteration is defined in regulation 3(2) of the Building Regulations 2010. An alteration is material for the purposes of the Building Regulations if the work, or any part of it, would at any stage result in (a) a building or controlled service or fitting not complying with a relevant requirement where previously it did; or (b) a building or controlled service or fitting which before the work commenced did not comply with a relevant requirement, being more unsatisfactory in relation to such a requirement.

**Material change of use (Building Regulations)**   The meaning of material change of use is defined in regulation 5 of the Building Regulations 2010. There is a material change of use where there is a change in the purposes for which, or the circumstances in which, a building is used. Note that a loft conversion in a single-family dwelling is not a material change of use. Material changes of use would include the conversion of a building into a dwelling (where previously it was not) and the conversion of a house into flats (where, for example, it was previously a single-family dwellinghouse).

**ODPM**   Office of the Deputy Prime Minister. Now defunct as a government department. See *Communities and Local Government*.

**OPC**   Ordinary Portland cement.

**OSB**   Oriented strand board. In loft conversions, principally used for external sheathing and roof decking.

**Outrigger**   See *back addition* (above).

**PD**   Permitted development.

**PFC**   Parallel flange channel (structural steel).

**PIR**   Polyisocyanurate (used in rigid thermal insulation boards).

**PUR**   Polyurethane (used in rigid thermal insulation boards).

**PWA/PWeA**   Party Wall etc. Act 1996.

**Parge coat**   A plaster or render coat used to seal bare masonry inside a building. It enhances thermal and sound performance, and improves airtightness.

**Party wall parapet**   Projection of the party wall above the plane of the roof between terraced or semi-detached dwellings, usually a minimum of 400 mm. Party wall parapets were once a regulatory requirement and were intended to inhibit the spread of fire between adjoining buildings. Widely used throughout the nineteenth century. See also *Gable parapet*.

**Plain tile**   Plain tiles are widely used as vertical cladding in loft conversions. Traditionally, plain tiles were made from clay, but concrete is now also used. The plain tile represents what is believed to be the earliest example of standardisation in English construction, its size set at 10½" × 6¼" by an enactment of 1477. Under the current standard, plain tiles are 265 × 165 mm.

**RdSAP**   Reduced data SAP (Standard Assessment Procedure). Official methodology for assessing the energy performance of existing dwellings. RdSAP is based on a survey of the property and is used when the complete data set for a full SAP calculation is not available.

**RMI**   Repair, maintenance and improvement (of buildings).

**RSJ**   Rolled steel joist. In strict terms, this is a designation in its own right (although seldom used now) and is distinct from the universal beam or column. However, it is widely used as a generic term to describe any steel I-beam.

**RWP**   Rainwater pipe (on drawings).

**SAP**   Standard Assessment Procedure. The government's official methodology for assessing the energy performance of dwellings.

**Sarking**   Timber boards or composite sheet material fixed to the outside of rafters on pitched roofs; not widely used in England and Wales.

**SEDBUK**   Seasonal Efficiency of Domestic Boilers in the UK.

**SVP**   Soil and vent pipe (on drawings).

**Simple payback**   For the purposes of Approved Document L1B, the amount of time it takes to recover the cost of energy-efficiency measures through energy savings actually made. Calculated by dividing the marginal additional cost of implementing an energy-efficiency measure by the value of the annual savings achieved through that measure, taking no account of VAT. For example, if the additional cost of implementing a measure (in materials and labour) was £600 and the value of the annual energy savings was £50 per year, the simple payback would be: 600/50 = 12 years.

**Skillings**   The name given to a sloping ceiling formed beneath a roof; for example, where the rafters of a pitched roof are boarded over to form a ceiling in a loft conversion. Sometimes called a skilling ceiling, skeiling or skeilings.

**Swept valley**   A curved valley formed in tile, slate or stone rather than lead or zinc.

**Torching**   In the roofs of older buildings, mortar applied to the underside of slates or tiles to prevent penetration of wind-blown rain and to resist wind-lift of cladding material. Rendered obsolete by the advent of building papers, felts and membranes.

**TRADA**   Timber Research and Development Association.

**Triangulation**   The creation of stable triangular configurations, particularly in roof structures. Distinct from triangulation in surveying/cartography.

**Underdraw**   To provide a layer, such as a ceiling, to the underside of a roof or floor.

**U-value**   The amount of heat energy (in watts) transmitted through $1\,m^2$ of building (e.g. a wall) for every $1°$ difference (K) between external and internal temperature – $W/m^2K$.

**VCL**   Vapour control layer (see below).

**Vapour check**   A vapour-impermeable membrane, often heavy-gauge polythene sheet with taped joints, used to prevent moisture-laden air from within a building escaping into and condensing within its structural fabric (see also *Interstitial condensation*). Sometimes referred to as a VCL or vapour barrier. Some plasterboards and rigid-sheet insulation material have an integral vapour check.

**WBP**   Plywood – weather and boil proof. Widely used in sheathing and decking. Sometimes described as WPB.

# Bibliography and useful contacts

## Primary references and standards

BRE (2009) *The government's standard assessment procedure for energy rating of dwellings (SAP 2009)*. Published on behalf of DECC.

BS 5250:2002 *Code of practice for control of condensation in buildings.*

BS 5588-1:1990 *Fire precautions in the design, construction and use of buildings. Code of practice for residential buildings.*

BS 5839-6:2004 *Fire detection and fire alarm systems for buildings. Code of practice for the design, installation and maintenance of fire detection and fire alarm systems in dwellings.*

BS 9251:2005 *Sprinkler systems for residential and domestic occupancies. Code of practice.*

BS EN 1990:2002 *Eurocode: Basis of structural design (Eurocode 0).*

BS EN 1991 *Eurocode 1: Actions on structures.*

BS EN 1993 *Eurocode 3: Design of steel structures.*

BS EN 1995 *Eurocode 5: Design of timber structures.*

BS EN ISO 13788:2002 *Hygrothermal performance of building components and building elements. Internal surface temperature to avoid critical surface humidity and interstitial condensation. Calculation methods.*

TRADA Technology Ltd (2009) *Eurocode 5 span tables for solid timber members in floors, ceilings and roofs for dwellings.*

## Publications and technical literature

Billington, M.J., Bright, K. and Waters, J.R. (2007) *The Building Regulations: Explained and Illustrated.* Wiley-Blackwell, Oxford.

BRE (1988) *Increasing the fire resistance of existing timber floors.* BRE Digest 208.

BRE (1998) *Timbers: Their Natural Durability and Resistance to Preservative Treatment.* BRE Digest 429.

BRE (2002) *Thermal insulation: avoiding risks.* BRE Report 262.

BRE (2006) *Conventions for U-value calculations.* BRE Report 443.

Brick Development Association (2001) *Observations on the use of reclaimed clay bricks.* PBM 1.4.

Brick Development Association (2005) *The BDA Guide to Successful Brickwork.* Elsevier.

Brunskill, R.W. (1985) *Timber Building in Britain.* Victor Gollancz Ltd, London.

Brunskill, R. & Clifton-Taylor, A. (1977) *English Brickwork.* Ward Lock Ltd, London.

Clay Roof Tile Council (2004) *A Guide to Plain Tiling Including Vertical Tiling.*

Cobb, F. (2004) *Structural Engineer's Pocket Book.* Elsevier.

Department for Communities and Local Government (2007) *Accredited construction details.*

Department for Communities and Local Government (2010) *Fire Statistics, United Kingdom 2008.*

Department for Communities and Local Government (2010) *Permitted development for householders – technical guidance.*

Department for Communities and Local Government (2011) *Guide to determinations and appeals under the Building Act 1984.*

English Heritage (2004) *Building Regulations and Historic Buildings.*

Goring, L. (2010) *Manual of First & Second Fixing Carpentry.* Elsevier.

Lead Sheet Association (2007) *Rolled Lead Sheet – The Complete Manual.*

Mindham, C.N. (2006) *Roof Construction and Loft Conversion.* Wiley-Blackwell, Oxford.

TRADA Technology Ltd (2001) *Timber Frame Construction.*

TRADA Technology Ltd (2005) *Fire Resisting Doorsets by Upgrading.* WIS 1–32.

Waste & Resources Action Programme (WRAP) *Reclaimed Building Products Guide.*

Yeomans, D. (1997) *Construction Since 1900: Materials.* Batsford, London.

## REGULATORY POLICY

**Building Regulations: England and Wales**
Sustainable Buildings Division
Department for Communities and Local Government
Eland House
Bressenden Place
London SW1E 5DU
Tel: 030 3444 0000
www.communities.gov.uk

**Building Regulations: Northern Ireland**
Building Regulations Unit
Department of Finance and Personnel
Level 5, Causeway Exchange
1–7 Bedford Street
Belfast BT2 7EG
Tel: 028 9051 2704
www.dfpni.gov.uk

**Building Regulations: Scotland**
Building Standards Division
Denholm House
Almondvale Business Park
Livingston EH54 6GA
Tel: 01506 600400
www.scotland.gov.uk/Topics/Built-Environment

## CONTACTS

**The Association of Building Engineers (ABE)**
Lutyens House
Billing Brook Road
Weston Favell
Northampton NN3 8NW
Tel: 0845 126 1058
www.abe.org.uk

**Association for the Conservation of Energy (ACE)**
Westgate House
2a Prebend Street
London N1 8PT
Tel: 020 7359 8000
www.ukace.org

**Brick Development Association Ltd (BDA)**
The Building Centre
26 Store Street
London WC1E 7BT
Tel: 020 7323 7030
www.brick.org.uk

**British Automatic Fire Sprinkler Association Ltd (BAFSA)**
Richmond House
Broad Street
Ely CB7 4AH
Tel: 01353 659187
www.bafsa.org.uk

**British Fenestration Rating Council (BFRC)**
54 Ayres Street
London
SE1 1EU
Tel: 020 7403 9200
www.bfrc.org

**British Standards Institution (BSI)**
389 Chiswick High Road
London W4 4AL
Tel: 020 8996 9001
www.bsigroup.com

**British Woodworking Federation (BWF)**
Royal London House
22–25 Finsbury Square
London EC2A 1DX
Tel: 0844 209 2610
www.bwf.org.uk

**Building Centre**
26 Store Street
London WC1E 7BT
Tel: 020 7692 4000
www.buildingcentre.co.uk

**Building Research Establishment (BRE)**
Bucknalls Lane
Garston
Watford WD25 9XX
Tel: 01923 664000
www.bre.co.uk

**Clay Roof Tile Council**
Federation House
Station Road
Stoke-on-Trent ST4 2SA

Tel: 01782 744631
www.clayroof.co.uk

**Construction Fixings Association**
65 Deans Street
Oakham LE15 6AF
Tel: 01664 823687
www.fixingscfa.co.uk

**Copper Development Association (CDA)**
5 Grovelands Business Centre
Boundary Way
Hemel Hempstead HP2 7TE
www.copperinfo.co.uk

**Corus** (see *Tata Steel*)

**Energy Saving Trust (EST)**
21 Dartmouth Street
London SW1H 9BP
Tel: 020 7222 0101
www.energysavingtrust.org.uk

**Federation of Master Builders (FMB)**
Gordon Fisher House
14–15 Great James Street
London WC1N 3DP
Tel: 020 7242 7583
www.fmb.org.uk

**Fire Protection Association (FPA)**
London Road
Moreton-in-Marsh
Gloucestershire GL56 0RH
Tel: 01608 812500
www.thefpa.co.uk

**Glued Laminated Timber Association (GLTA)**
Chiltern House
Stocking Lane
High Wycombe HP14 4ND
Tel: 01494 565180
www.glulam.co.uk

**Institution of Structural Engineers**
11 Upper Belgrave Street
London SW1X 8BH
Tel: 020 7235 4535
www.istructe.org.uk

**LABC**
Third Floor
66 South Lambeth Road

London SW8 1RL
Tel: 020 7091 6860
www.labc.uk.com

**Lead Sheet Association (LSA)**
Unit 10, Archers Park
Branbridges Road
East Peckham
Tonbridge TN12 5HP
Tel: 01622 872432
www.leadsheetassociation.org.uk

**NBS (National Building Specification)**
The Old Post Office
St. Nicholas Street
Newcastle upon Tyne NE1 1RH
Tel: 0191 244 5500
www.thenbs.com

**Pyramus & Thisbe Club**
Rathdale House
30 Back Road
Rathfriland BT34 5QF
Tel: 028 4063 2082
www.partywalls.org.uk

**Royal Institute of British Architects (RIBA)**
66 Portland Place
London W1B 1AD
Tel: 020 7580 5533
www.architecture.com

**Royal Institution of Chartered Surveyors (RICS)**

Parliament Square
London SW1P 3AD
Tel: 0870 333 1600
www.rics.org

**Steel Construction Institute (SCI)**
Silwood Park
Ascot SL5 7QN
Tel: 01344 636525
www.steel-sci.org

**Tata Steel**
Construction Services and Development
PO Box 1
Brigg Road
Scunthorpe DN16 1BP
Tel: 01724 405060
www.tatasteelconstruction.com

**Timber Research and Development Association (TRADA)**
Stocking Lane
Hughenden Valley
High Wycombe
Buckinghamshire HP14 4ND
Tel: 01494 569600
www.trada.co.uk

**Trussed Rafter Association (TRA)**
The Building Centre
26 Store Street
London WC1E 7BT
Tel: 020 3205 0032
www.tra.org.uk

# ONLINE APPLICATIONS FOR PLANNING AND BUILDING CONTROL

**Submit-a-Plan**
Building control applications for England, Wales and Northern Ireland.
www.submitaplan.com

**Planning Portal**
Planning applications for England and Wales.
www.planningportal.gov.uk

# Index

*Loft Conversions*, Second Edition. John Coutts.
© 2013 John Coutts. Published 2013 by Blackwell Publishing Ltd.

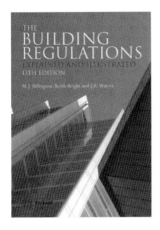
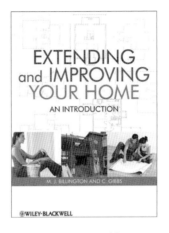
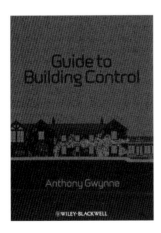
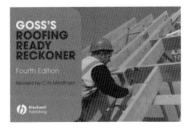

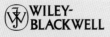